GeoPlanet: Earth and Planetary Sciences

Series Editors

Paweł Rowiński (Editor-in-Chief)
Marek Banaszkiewicz
Janusz Pempkowiak
Marek Lewandowski

For further volumes:
http://www.springer.com/series/8821

Adam F. Idziak • Ryszard Dubiel

Editors

Geophysics in Mining and Environmental Protection

 Springer

Editors
Adam F. Idziak
University of Silesia
Department of Applied Geology
Będzińska 60
41-200 Sosnowiec
Poland
adam.idziak@us.edu.pl

Ryszard Dubiel
University of Silesia
Department of Applied Geology
Będzińska 60
41-200 Sosnowiec
Poland
ryszard.dubiel@us.edu.pl

The GeoPlanet: Earth and Planetary Sciences Book Series is in part a continuation of Monographic Volumes of Publications of the Institute of Geophysics, Polish Academy of Sciences, the journal published since 1962 (http://pub.igf.edu.pl/index.php).

ISSN 2190-5193   e-ISSN 2190-5207
ISBN 978-3-642-27040-6   ISBN 978-3-642-19097-1 (eBook)
DOI 10.1007/978-3-642-19097-1
Springer Heidelberg Dordrecht London New York

Cover design: deblik, Berlin

Printed on acid-free paper

Springer is part of Springer Science+Business Media (www.springer.com)

Series Editors

Geophysics: Paweł Rowiński
Editor in-Chief
Institute of Geophysics
Polish Academy of Sciences
Ks. Janusza 64
01-452 Warszawa, Poland
p.rowinski@igf.edu.pl

Space Sciences: Marek Banaszkiewicz
Space Research Centre
Polish Academy of Sciences
ul. Bartycka 18A
00-716 Warszawa, Poland

Oceanology: Janusz Pempkowiak
Institute of Oceanology
Polish Academy of Sciences
Powstańców Warszawy 55
81-712 Sopot, Poland

Geology: Marek Lewandowski
Institute of Geological Sciences
Polish Academy of Sciences
ul. Twarda 51/55
00-818 Warszawa, Poland

Managing Editor

Anna Dziembowska
Institute of Geophysics, Polish Academy of Sciences

Advisory Board

Tilman SPOHN
Institut für Planetenforschung
Deutsches Zentrum für Luft- und Raumfahrt
in der Helmholtz Gemeinschaft
Berlin, Germany

Krzysztof STASIEWICZ
Swedish Institute of Space Physics
Uppsala, Sweden

Roman TEISSEYRE
Earth's Interior Dynamics Lab.
Institute of Geophysics,
Polish Academy of Sciences
Warszawa, Poland

Jacek TRONCZYNSKI
Laboratory of Biogeochemistry
of Organic Contaminants,
IFREMER DCN_BE
Nantes, France

Steve WALLIS
School of the Built Environment,
Heriot-Watt University, Riccarton,
Edinburgh, Scotland, UK

Wacław M. ZUBEREK
Department of Applied Geology
University of Silesia
Sosnowiec, Poland

Preface

Symposia of Polish, Czech and Slovakian scientists and practitioners engaged in geophysical research in mines have a long history. Joint meetings have been organized regularly since 1970, alternately in Poland, Czechoslovakia and then the Czech Republic. Being initially devoted to Mining Geophysics, their scope was expanded in 2002 to include broader environmental problems. Outcomes of the Symposia held in Poland were published in a series of books within the series *Publications of the Institute of Geophysics, Polish Academy of Sciences*.

This volume is an outcome of the 32nd Polish-Czech-Slovakian Symposium on Mining and Environmental Geophysics which was held in Piechowice (the Sudety Mts., Southern Poland) on May 19–22, 2009. The Symposium was organized by the Faculty of Earth Sciences, University of Silesia, and the Institute of Geophysics, Polish Academy of Sciences, under the honorary auspices of the President of Polish State Mining Authority. The Symposium assembled 56 participants who presented 35 papers.

The main topics of the 32nd Symposium were the following:

- Mining and natural seismicity
- Mining geophysical methods
- Application of geophysics in environmental studies and geology

The goal of the Symposium was to present and discuss the results of current research related to mining and environmental geophysics. The Symposium was also a good and convenient opportunity for discussion and exchange of various concepts among participants representing universities, research centers and industrial institutions. The Symposium proceeded in nice and friendly atmosphere. We would like to express our sincere thanks and appreciation to the members of Scientific and Local Organizing Committees.

The volume contains thirteen papers selected from those presented at the 32nd Polish-Czech-Slovakian Symposium. The papers were peer reviewed and then revised according to reviewers' comments and suggestions. Finally the papers were accepted for publication.

Representing the wide range of Symposium topics, the papers included into this volume touch multiple aspects of geophysical research. You can find papers related to seismology (mining and global), geodesy, environmental geophysics and geology alike.

We hope that the readers will find the volume interesting and useful.

Sosnowiec, Poland   Adam F. Idziak
                                                                                     Ryszard Dubiel

Contents

Chapter 1
Evaluation of Seismic Effects Generated by Mining Induced Seismic Events in Stonava Area

Zdeněk Kaláb and Jaromír Knejzlík

Abstract Stonava area represents a region with high seismic loading due to underground mining. For this reason, several seismological measurements are taken in buildings on the surface. This paper describes the results of the measurements taken in November 2008 and spring 2009 when experimental seismic stations were operated in the region under discussion. The area selected for measurement is not large enough for evaluating the validity of data measured at a certain point for its surroundings. A comparison study of recorded values on individual experimental seismic stations is presented in this paper.

1.1 Introduction

Long-term underground mining of mineral resources can provoke, among other things, the origin of mining induced seismicity due to complicated and unfavorable geological conditions. The region of Karviná (North Moravia, Czech Republic) is a typical example, because more than 100 years of underground mining of black coal causes a complex structure of mined-out and buried spaces, unworked layers and safety pillars. These facts together with regional geological and geomechanical situation result not only in seismic events at stope faces but also induce very intensive seismic events that implicate macroseismic effects in workings and on the surface

Z. Kaláb (✉)
Institute of Geonics of the Academy of Sciences of the Czech Republic, v.v.i., Studentská 1768, 708 00 Ostrava-Poruba, Czech Republic
and
VŠB-Technical University of Ostrava, 17. Listopadu 15, 708 33 Ostrava-Poruba, Czech Republic
e-mail: kalab@ugn.cas.cz

J. Knejzlík
Institute of Geonics of the Academy of Sciences of the Czech Republic, v.v.i., Studentská 1768, 708 00 Ostrava-Poruba, Czech Republic
e-mail: knejzlik@ugn.cas.cz

A.F. Idziak and R. Dubiel (eds.), *Geophysics in Mining and Environmental Protection*, Geoplanet: Earth and Planetary Sciences 2, DOI 10.1007/978-3-642-19097-1_1,
© Springer-Verlag Berlin Heidelberg 2011

(mine tremors, rockbursts). Damping program of exploitation in the Karviná region contributes to selective exploitation due to economical reasons. At this time, adequate reduction of the number and intensity of seismic events is not observed in the Karviná region. Current exploitation provokes higher load of mining fields and complicated geometry of worked-out spaces. Exploitation in deeper seams and the necessity of exploitation of residual coal parts in complicated conditions on contact areas between mined and mined-out spaces are other reasons for current seismic loading. It is obvious that the intensity of mining induced seismicity is a result of combination of current and previous mining activities (Konečný et al. 2003).

Generally, local geological pattern influences values of seismic vibrations on buildings and structures, i.e., the site effect (e.g., Bard 2008; Pitilakis 2007; Ansal 2004; Olszewska and Lasocki 2004). This is also documented by our results that are derived from previous experimental seismological measurements in the Karviná region (e.g., Doležalová et al. 2008; Kaláb and Lyubushin 2008; Kaláb and Knejzlík 2002, 2006, 2008; Holečko et al. 2006). Seismic impacts on buildings in the most endangered parts of Karviná region are monitored using solitaire seismic stations operated by the Institute of Geonics (only five stations in the 10×7 km area). Results from our measurements in surface structures document that the most intensive seismic events exceed value of 10 mm s^{-1} (component velocity; component accelerations reach 500 mm s^{-2}). Seismic effects of the most intensive mining induced seismic events on buildings have higher values in small epicenter distances; however, limit values for more significant damages (i.e., damage degree No. 2 according to Czech Technical Standard 73 0040) are not exceeded for most of the measured cases. It is necessary to take into account that extremely intensive seismic event has not occurred during monitoring periods. Measured values for residential buildings (classes B and C, according to Standard 73 0040) correspond with damage degree 0 ("without damages") or degree 1 ("first cosmetic damages"); only for buildings of class A (e.g., historical buildings) and in the most unfavorable ground conditions, damage degree No. 2 ("minor damages") is obtained.

It is not documented if measured velocity values are influenced by the site effect and whether or not these values are valid for other buildings in the surroundings. The situation in which all endangered buildings should have been monitored cannot be achieved for technical and economical reasons. Study of seismic loadings in the model region depending on local geological conditions is the main task of this contribution.

1.2   Experimental Seismological Measurement

Results from mathematical modeling and simulation will be integral parts of all interpretations. First calculations for Karviná region using finite element method were aimed at modeling of the geological situation changes on character of the seismic record (Hrubešová and Kaláb 2005; Kaláb et al. 2008). These planar calculations, using program system Plaxis 2D, show suitability of this procedure.

To have more precise information about surface wave field, short-term seismological measurements were taken in autumn 2008 and spring 2009. The Stonava area, eastern part of the Karviná region, was selected for this experiment.

The measurements have been taken with a different type of apparatuses to obtain utmost data. Seismic recording apparatuses of the PCM3-EPC2 type that were developed in the Institute of Geonics are prepared for measurements (sampling frequency of 100 or 250 Hz, dynamic range of 90 dB). These instrumentations are used for permanent monitoring on solitaire stations. We also used GAIA seismic recorders produced by Vistec Company (sampling frequency up to 500 Hz, dynamic range of 128 dB). It is possible to operate these recorders with different types of seismometers (i.e., Le3D, GSV-320 and ViGeo 2).

Two permanent seismic stations (named STO1 and STO2) are operated in the Stonava area. Experimental measurement in the selected area was taken on four temporary stations between 7 October and 24 October, 2008. These points were labeled KAR A, KAR B, KAR C and KAR D (Fig. 1.1). All points were placed in cellars of buildings. The distance between points KAR A and KAR C was less than 1 km, the other distances were smaller. Temporary stations were operated in continuous regime of records (stations KAR D operated 3 days only due to a technical problem). Sampling frequency of digital data was 250 Hz and frequency range of seismic channel was 2–80 Hz. Localities in which temporary seismic stations were

Fig. 1.1 Distribution of seismic stations in Stonava area during experiment in autumn 2008 (base: www.mapy.cz) and thematic layer map called "seismic characteristics of foundation conditions"; base map: A=KAR A, B=KAR B, C=KAR C, D=KAR D, 1=STO1, 2=STO2

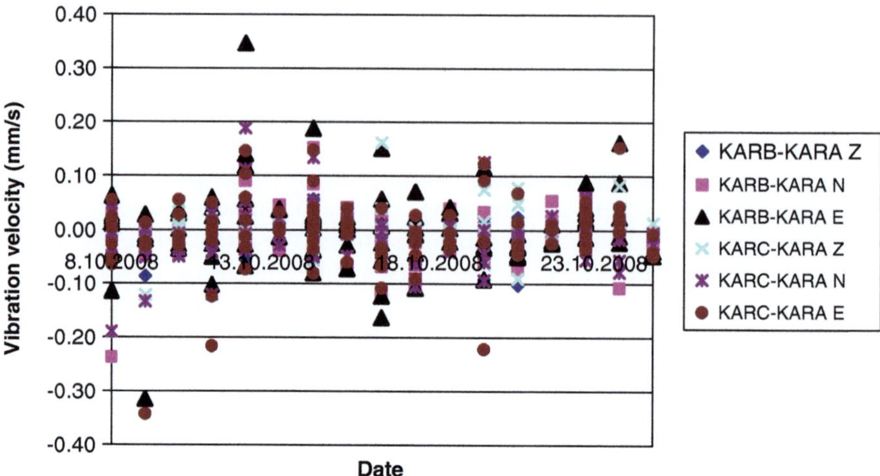

Fig. 1.2 Calculated differences between individual stations (values from station KAR A are the base; Z means the vertical component, N the N–S component, and E the E–W component)

placed can be included into places with unfavourable and/or critical class of foundation conditions (Lednická and Kaláb 2008a) using geological conditions as criterion. Map of thematic layer called "seismic characteristic of foundation conditions" for Stonava area is part of Fig. 1.1 (Lednická and Kaláb 2008b). Therefore, an increased seismic effect of induced vibrations can be expected.

More than 100 mining induced seismic events were detected from continuous three-component records. Duration of this measurement was short and no more intensive mining induced seismic events occurred during this time. The maximum component velocity with value of 0.62 mm s^{-1} was obtained on 9 October 2008 (00:28 local time) event. The maximum value of absolute differences between individual stations was 0.34 mm s^{-1} (Fig. 1.2), this maximum difference value is almost the same for all three fixed bases (stations). More detailed description of this experiment was presented by Kaláb et al. (2009).

1.3 Second Experimental Seismological Measurement

As mentioned above, no significant macroseismic effect occurred during first experimental measurements. Significant effects were documented in November and December 2008 when two very strong events occurred. These shocks evoked very intensive vibrations on the surface; the maximum value of component velocity exceeded 20 mm s^{-1} (measurement from permanent seismic station). Second experimental measurement was again focused on evaluation of seismic loading of existing buildings and mutual differences. It was started since 2 April 2009, and ended on 4 June 2009. Five temporary seismic stations located in buildings were

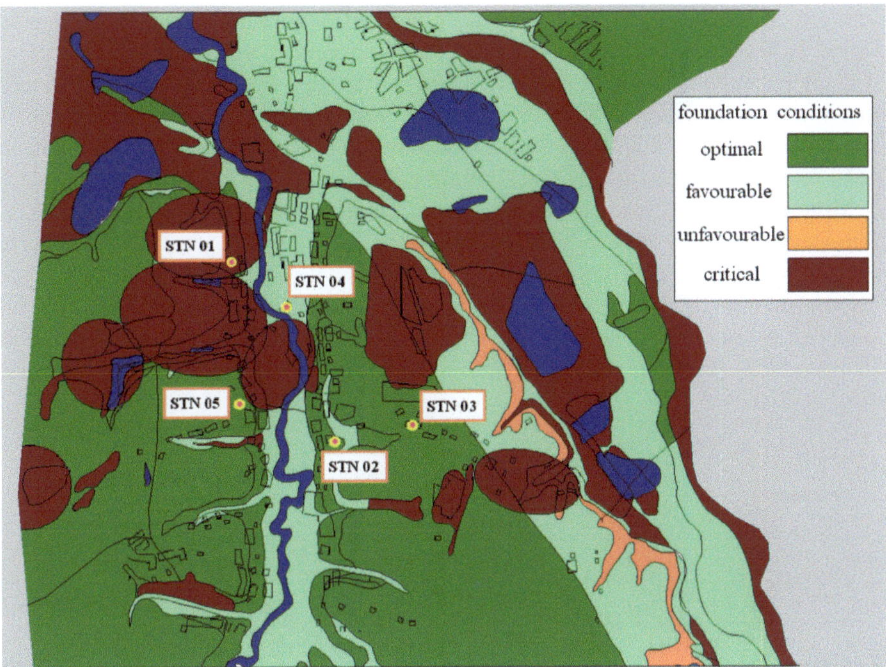

Fig. 1.3 Thematic layer map called "seismic characteristics of foundation conditions" and distribution of seismic stations in Stonava area during experiment in spring 2009

used during this experiment; the distribution of stations is presented in Fig. 1.3. Seismic stations KAR A, STO1 and STO2 were used again and they were re-named STN 01, STN 04 and STN 05, respectively. Stations STN 02 and STN 03 were placed into new positions. Foundation conditions for seismic stations in this experiment were not the same: three points had optimal conditions, one point had favorable conditions and one point had critical conditions. The distance between points STN 01 and STN 03 was about 2.5 km, other distances were smaller.

About 450 mining induced seismic events were recorded during the second experiment. Events with component amplitude greater than 0.1 mm s^{-1} were taken into consideration for interpretation. This set is represented by 50 events that occurred inside the experimental seismic network or in near surroundings. All these events have seismic energy up to 10^5 J (according to the classification made by Green Gas Paskov). Only one seismic event occurred outside the network; about 8 km to WNW from centre of the network. Seismic energy of this event was about 8×10^5 J, local magnitude 1.9 (according to the classification made by Green Gas Paskov). This event is interpreted separately.

The most intensive seismic velocities were detected on N-S components. Therefore, the next figures will represent this component only. The maximum velocity component value was almost 2 mm s^{-1}; this is a very small value from the seismic loading of building and structures. The highest values of velocity were frequently detected on station STN 01 (station with critical foundation conditions). Seismic station STN 04 was selected as the base. This station is approximately in the centre of the experimental network and a permanent station is operated there. The maximum value of absolute differences between individual stations was 1.61 mm s^{-1} (Fig. 1.4). Relative differences between individual stations are unusable due to small recorded values, up to several thousand percent.

Mining induced seismic events that occurred as a consequence of coal exploitation seem to be directly under the base of the experimental network; they were selected for second set. This set contains 12 events with seismic energy in the range 2×10^3–6.4×10^4 J, depths of events were about 600 m (according to the classification made by Green Gas Paskov). The maximum velocity component value was 1.7 mm s^{-1}; the maximum value of absolute differences between individual stations was 1.61 mm s^{-1} (Fig. 1.5).

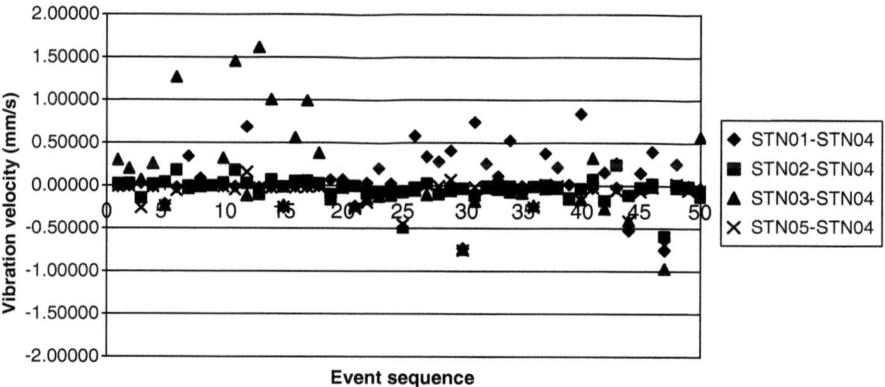

Fig. 1.4 Calculated differences between individual stations, set of 50 mining induced seismic events, N component (values from station STN 04 are base)

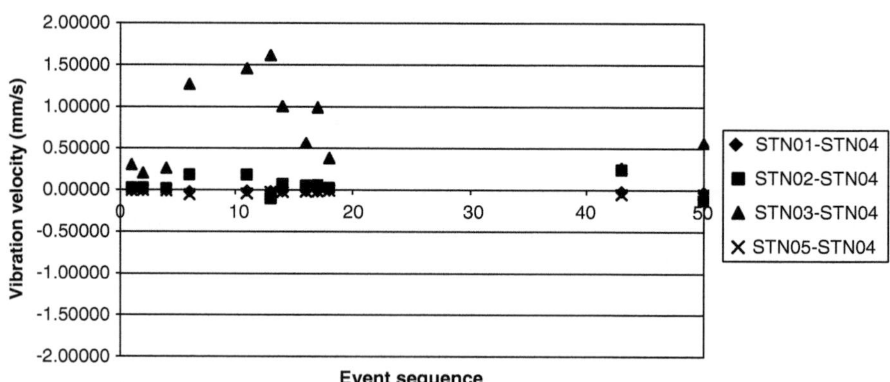

Fig. 1.5 Calculated differences between individual stations, set of 12 mining induced seismic events occurred in the base of the experimental network, N component (values from station STN 04 are the base)

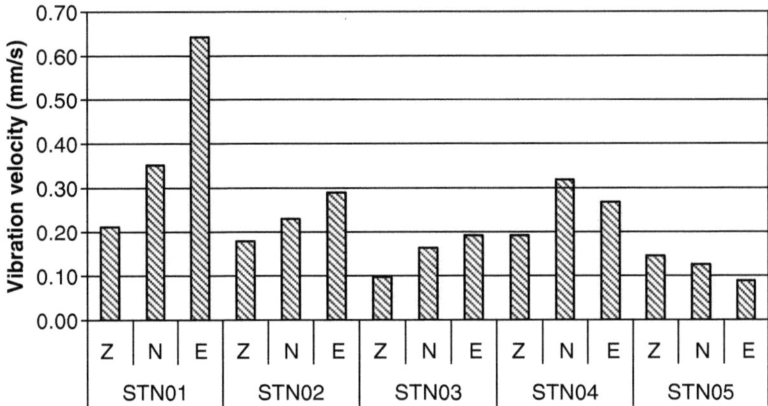

Fig. 1.6 Measured component vibration velocities on individual seismic stations for 26 April 2009 mining induced seismic event

As mentioned above, only one more intensive mining induced seismic event occurred during experimental measurements (26 April 2009, 02:51 local time). Its basic parameters were described above. Figure 1.6 presents measured component vibration velocities on individual seismic stations. All values are practically the same, except those measured on STN 01 seismic station. The first reason is that STN 01 is the nearest station but dimension of the network is not so high. Therefore, it is possible to derive that critical foundation conditions are responsible for these higher values.

1.4   Conclusions

To determine values of velocity vibration generated by mining induced seismic events on the surface, seismological parameters of events and local geology must be taken into account. Theoretical solution of site effect interpretation was published by many authors. However, the majority of these recommendations appear from "simple" geological conditions and strong seismic events. For example, Haghshenas et al. (2008) wrote that "horizontal-to-vertical (H/V) method can provide a very good estimation of fundamental frequency of soil sites. However, there are a few cases where H/V microtremors ratio fails to predict the correct frequency. The H/V technique works very well on soft deposits underlain by hard bedrock or much stuffer sediments, and is less efficient on stiff, thick deposits. The H/V method cannot give a good estimation of the amplification value at a site. H/V peak amplitudes and spectral ratio method amplifications are not correlated and highly scattered" (for details see the cited paper).

Local geology in Stonava area is very complicated: a thin layer of soft Quaternary sediments that covered the Tertiary sediments of different thickness (from 300 up to 600 m) and different stiffness. In the area under study, several significant faults are found. Using mining induced seismic events and ambient noise for evaluation of site effect documents that there exist points with significant differences in the H/V spectral ratio curves and points with the same shapes of the H/V spectral ratio curves. Significant sources of vibrations in the surroundings of the seismic station are a probable reason of this effect. Industrial seismicity (factory, forging shop, mining ventilation, etc.) has significant sense in whole Karviná region (Kaláb and Knejzlík 2008; Kaláb and Lyubushin 2008).

The study of seismic loadings in Stonava area depending on local geological conditions was the main task of experimental measurements. Basic conclusions from experiments can be summarized as follows:

- Weak seismic events that occurred within the base of the experimental network have significant differences between registrations on individual seismic stations. However, the component velocities obtained are small, concerning seismic loading of buildings.
- More intensive seismic events did not occur during experimental measurements; therefore, it is impossible to derive recommendations.
- For weak and more intensive seismic events from the surroundings, seismic loading of individual places is quite comparable.
- The strongest far event has comparable component velocities, except for a seismic station with critical foundation conditions. The highest values were recorded there.

Our experimental measurements document that an area of about 2×2 km needs at least two seismic monitoring places, one in favourable foundation conditions and the other in critical foundation conditions.

Acknowledgement This paper has been achieved with the financial support of the Czech Science Foundation, project No. 105/07/0878.

References

Ansal, A. (2004) Recent Advances in Earthquake Geotechnical Engineering and Microzonation. Kluwer Academic Publisher, Dordrecht, 353 pages.

Bard, P-Y. editor (2008) The H/V Technique: Results of the SESAME Project. Bulletin of Earthquake Engineering, Vol. 6, No. 1, Special Issue, 147 pages.

Doležalová, H., Holub, K. and Kaláb, Z. (2008) Underground Coal Mining in the Karviná Region and Its Impact on the Human Environment (Czech Republic). Moravian Geographical Report, Vol. 16, No. 2, 14-24.

Haghshenas, E., Bard, P.-Y. and Theodulidis, N. (2008) Empirical Evaluation of Microtremor H/V Spectral Ratio. Bull. Earthquake Eng. Vol. 6, No. 1, 75-108.

Holečko J., Kaláb, Z., Knejzlík J. and Ptáček, J. (2006) Vibration Velocity on the Surface in Karviná Region of the Upper Silesian Coal Basin. Journal Coal-Ores-Geological Survey, No. 2/2006, Czech Rep., 34-39.

Hrubešová, E. and Kaláb, Z. (2005) Example of Modeling of the Mining Induced Seismicity Impact on the Building Using Numerical System Plaxis. In: Konečný, P. (ed): EUROCK 2005 – Impact of Human Activity on the Geological Environment, A.A. Balkema Publisher, Leiden, 213-218.

Kaláb Z. and Knejzlík J. (2002) Systematic Measurement and Preliminary Evaluation of Seismic Vibrations Provoked by Mining Induced Seismicity in Karviná Area, Publs. Inst. Geophys. Pol. Acad. Sc., M-24(340), 95-103.

Kaláb, Z. and Knejzlík, J. (2006) Field Measurement of Surface Seismic Vibrations Provoked by Mining in Karvina Region. Publs. Inst. Geophys. Pol. Acad. Sc., 29(395), Warszawa, 185-194.

Kaláb, Z. and Knejzlík, J. (2008) Spectral Ratio Evaluation of Mining Induced Seismic Events from Karviná Region (Czech Republic). Proceedings of European Seismological Commission ESC2008, 31st General Assembly, CD – Short Papers electronic volume, 175-181.

Kaláb, Z. and Lyubushin, A.A. (2008) Study of Site Effect using Mining Induced Seismic Events and Ambient Noise from Karviná Region. Acta Geodyn. et Geomater., Vol. 5, No. 2(150), ISSN 1214-9705, 105-113.

Kaláb Z., Knejzlík J. and Hrubešová E. (2008) Influence of Local Geology on Ground Velocity on the Surface in Karviná Area. Journal Coal-Ores-Geological Survey, No. 1/2008, 26-31 (in Czech).

Kaláb, Z., Knejzlík, J. and Lednická, M. (2009) Seismological Experiment in Stonava Area: Evaluation of Validity of Measured Data. Górnictwo i Geologia, Kwartalnik, Tom 4, Zeszyt 2a, Wydawnictwo Politechniki Slaskiej, Gliwice, Poland, 95-103.

Konečný, P., Velička, V., Šňupárek, R., Takla, G. and Ptáček, J. (2003) Rockbursts in the Period of Mining Activity Reduction in Ostrava – Karviná Coalfield. ISRM 2003 – Technology Roadmap for Rock Mechanics, South African Institute of Mining and Metallurgy.

Lednická, M. and Kaláb, Z. (2008a) Maps of Clash of Opinions – Classification of Geological Conditions in Karviná Region. Zeszyty Naukowe Politechniki Śląskiej, Ser. Górnictwo z. 283, Nr. 1781, Gliwice, 139-147.

Lednická, M. and Kaláb, Z. (2008b) Pilot Map of New Thematic Layer Called "Seismic Characteristic of Foundation Conditions" for Selected Area of Karviná Region. 70 Years of FCE STU – Proceedings of the International Conference, CD, Slovak University of Technology in Bratislava, Faculty of Civil Engineering, 9 pages (in Czech).

Olszewska D. and Lasocki S. (2004) Application of the Horizontal to Vertical Ratio Technique for Estimating the Site Characteristics of Ground Motion caused by Mining Induced Seismic Events. Acta Geophysica Polonica. Vol. 52, No. 3, 302-318.

Pitilakis, K.D.-ed. (2007) Earthquake Geotechnical Engineering. Proceedings of 4th Int. Conf., Springer, 487 pages.

Chapter 2
Moment Tensor Solution and Physical Parameters of Selected Recent Seismic Events at Rudna Copper Mine

Grzegorz Lizurek and Paweł Wiejacz

Abstract The Lubin Copper Basin is a source of induced seismic events of considerable size. The paper presents the source mechanism study of fifty seismic events from Rudna Copper Mine that occurred in 2007. The energies of studied seismic events were greater than 10^5 J. Focal mechanisms for those events have been calculated with seismic moment tensor inversion method using FOCI computer software. The calculations were made using L1 and L2 norms. Final results are presented for L1 norm due to its lower sensitivity to possible errors. Spectral analysis of P and S waves was also performed. The analysis shows that typical focal radius is of about 200 m.

2.1 Introduction

Seismic moment tensor solutions and physical parameters of seismic events in mines could be an extension of seismic event location and energy calculation routines in mines. The information about focal mechanisms and source parameters can be used in mining engineering together with the focal coordinates and energy of tremors that are used now. Estimation of the fault planes, focal radii and their spatial analysis done periodically can point out if any of direction in space is more often observed than others. If so, then it may prove useful to try to connect it to geological or mining features such as an old fault in or near mine region (re) activated during mining processes or new fault formation, the direction of caving works or any prevailing direction of cracks or layering in the rockmass. The knowledge of mechanisms of events and focal radius can be also taken into account in blasting procedure used in mining practice of stress release and extraction of

G. Lizurek (✉) • P. Wiejacz
Institute of Geophysics, Polish Academy of Sciences, Księcia Janusza 64, 01-452 Warszawa, Poland
e-mail: lizurek@igf.edu.pl

A.F. Idziak and R. Dubiel (eds.), *Geophysics in Mining and Environmental Protection*,
Geoplanet: Earth and Planetary Sciences 2, DOI 10.1007/978-3-642-19097-1_2,
© Springer-Verlag Berlin Heidelberg 2011

deposits. For the purpose of this study, we have selected 50 seismic events induced at Rudna Copper Mine in the second half of 2007. For those events we have calculated source parameters and source mechanisms with seismic moment tensor inversion method.

2.2 Rudna Copper Mine: Network and Data

Rudna Copper Mine is located in the Legnica-Glogow (Lubin) Copper District, Southwestern Poland. Seismograms of 50 mine induced events in 2007 of energy higher than 10^5 J recorded by Rudna Copper Mine seismological network (Fig. 2.1) were analyzed. Data selected for our study was provided by Rudna Copper Mine.

Spectral analysis of P and S waves, physical properties of foci and moment tensor solutions of all selected events were determined. Coordinates of foci and origin time were provided by Rudna Copper Mine. Due to huge errors of the Z coordinate estimation in near-plain geometry of mine network, event depth is commonly fixed at the excavation level. The Rudna Copper Mine local network is composed of 32 vertical seismometers at mining level, except of two sensors placed in elevator shafts. However, the sensors in shafts were very rarely used, because of the noise coming from the shafts. The depth range of seismometers location is from 300 down to 1,000 m below surface.

During the period of the study the most active mining divisions had about 15 events of energy higher than 10^5 J. Distribution of events in mine subdivisions according to their energy is shown in Fig. 2.2.

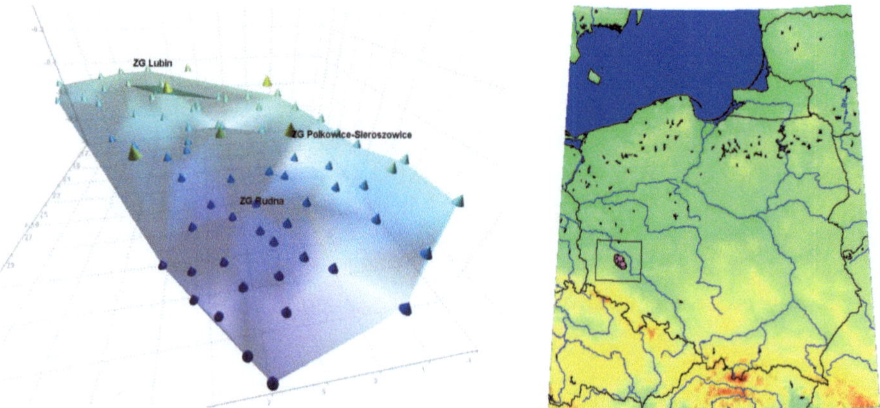

Fig. 2.1 The Rudna Copper Mine seismic network along with networks of neighbouring mines. *Right*: location of the area on the map of Poland

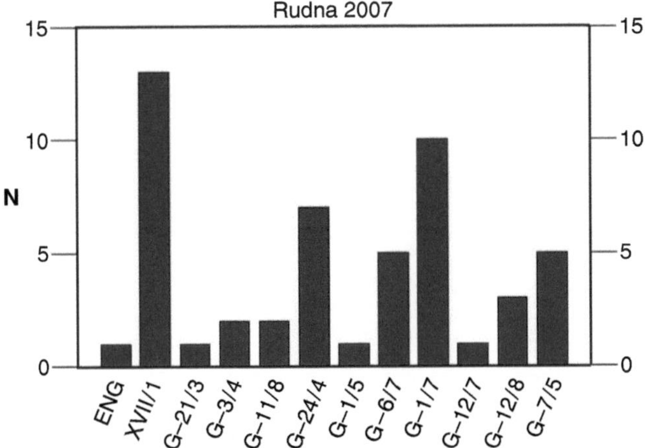

Fig. 2.2 Histogram of number of events in mine subdivisions

2.3 Spectral and Physical Parameters of Events

The basis of the study are ground motion spectra that had been obtained by integration of velocity seismograms. The relevant parts of seismograms of P and S waves were selected manually, and then transformed by Fast Fourier Transformation (FFT). Resulting amplitude spectra were corrected for attenuation effects with $Q = 400$ and 200, respectively, for P and S waves. For further calculations we set velocities of P and S waves at $Vp = 5{,}700$ m/s and $Vs = 3{,}300$ m/s. Far field displacement spectra were approximated on logarithmic scale. The two asymptotes were calculated using integrals of squared displacement and velocities of ground motion. This has allowed to estimate the spectral plateau, corner frequency and energy flux in observation point for every spectrum of P and S waves. Next, energies of P and S waves were calculated from the energy flux in far field displacement.

The observed S wave to P wave energy ratio is within the range from 6 to 70, but as it is shown in Fig. 2.3 the ratio value near 10 is typical to most of the events. Higher values of S to P wave energy ratio (higher than 20) indicate that in the focal mechanism the DC (Double-Couple) forces are dominant. The value of the ratio lower than 20 indicates that other components of the mechanism solution are also present.

Average seismic moment M_0, according to Aki and Richards (1980), $M_0 = \frac{4\pi\rho_0 c_0^3 R\Omega_0}{F_C R_C S_C}$, from all channels recorded for particular events was estimated with spectral plateau of P and S waves frequency spectra:

$$\Omega_0 = 2\left(\frac{K^3}{J}\right)^{\frac{1}{4}},$$

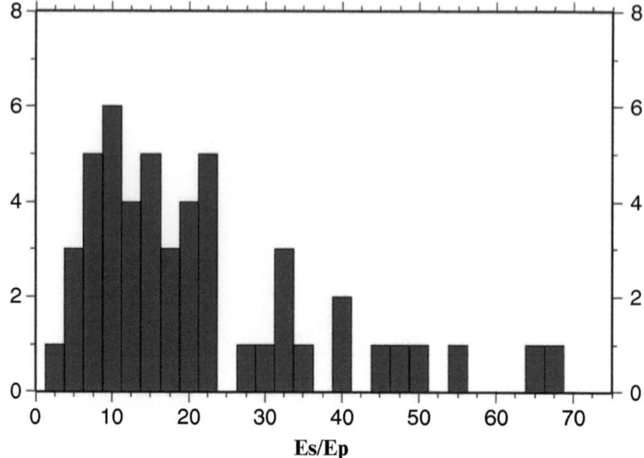

Fig. 2.3 Es vs. Ep ratio for analyzed events from Rudna mine

where J and K are as follows:

$$J = 2 \int_0^\infty |V(\omega)|^2 df = 2 \int_0^\infty \omega |U(\omega)|^2 df$$

$$= \frac{2}{3}[\Omega_0 \omega_1]^2 f_1 + 2 \int_{f_1}^{f_2} \omega |U(\omega)|^2 df + 2|\omega_2 U(\omega_2)|^2 f_2$$

$$K = 2 \int_0^\infty |U(\omega)|^2 df$$

$$= 2|U(\omega_1)|^2 f_1 + 2 \int_{f_1}^{f_2} |U(\omega)|^2 df + \frac{2}{3}|U(\omega_2)|^2 f_2$$

$$K = \frac{1}{4}\Omega_0^2 (2\pi f_C)$$

In these equations, $U(\omega)$ is the far-field displacement in the frequency domain, $V(\omega)$ is the far-field ground velocity in the frequency domain, f_1, f_2 are limits of the bandwidth of seismic instruments, ρ_0 is the density of source material, c_0 is the P or S-wave velocity at the source, R is the source-receiver distance, F_c accounts for radiation of P or S-waves, R_C accounts for the free surface amplification of either P or S-waves amplitudes, and S_C is the site correction for either P or S-waves (Snoke 1987; Andrews 1986).

Then the magnitude was calculated from the equation (Hanks and Kanamori 1979):

$$M_L = \frac{2}{3} \log M_0 - 6.05$$

The magnitude distribution of the events and seismic moment histogram are presented in Figs. 2.4 and 2.5, respectively. The analyzed events are characterized by narrow range of seismic moment values – 2.1×10^{12}–4.2×10^{14}, corresponding to magnitudes M from 2.2 to 3.7. Most of the tremors had smaller magnitudes (less than 2.8).

The focal radius R has been calculated using the formula of Madariaga (1976):

$$r_0 = \frac{K_C \beta_0}{2\pi f_C},$$

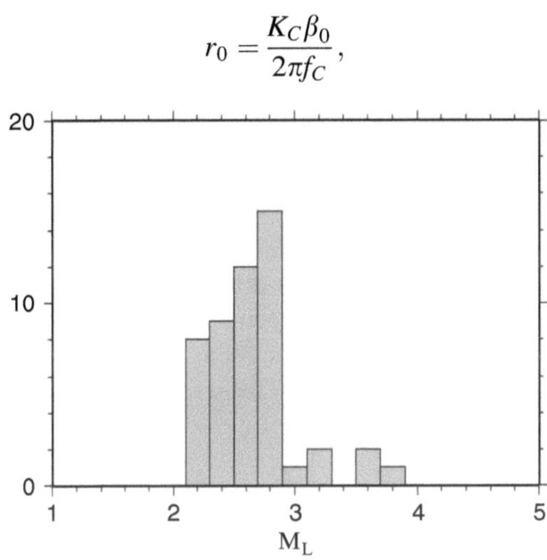

Fig. 2.4 Number of events of different magnitudes

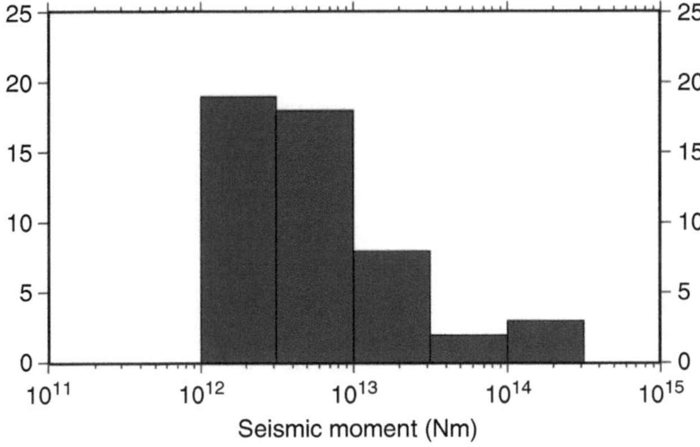

Fig. 2.5 Seismic moment histogram

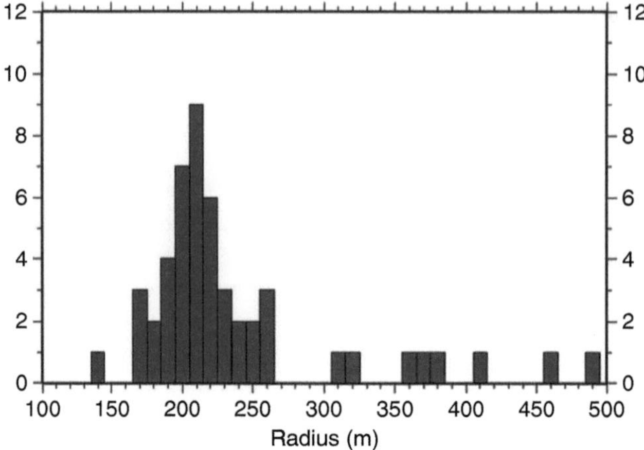

Fig. 2.6 Distribution of focal radii of studied events

where β_0 is the S-wave velocity in the source area, and K_C is a constant depending on the source model.

The results shown in Fig. 2.6 are mean values of radius calculated from all channels for a given event. The dominant value of focal radius appears to be 200 m, but the range of estimated radii is between 135 and 488 m. Therefore, it can be assumed that radius size region around the focus is weakened by the rupture cracks that evolved during the seismic event. The size of focal area is considerable as compared to the gallery size in the mine.

2.4 Focal Mechanism: Moment Tensor Solutions

Focal mechanisms were calculated by moment tensor inversion technique with use of first arrivals, amplitudes and signs of the arrivals. Norms L1 and L2 were applied to calculations; the depth of focus was assumed as an excavation level. All analyses and calculations were done with FOCI software.

Three types of seismic moment tensor have been calculated for each event: full moment tensor, trace null tensor and double couple (DC) moment tensor. The full moment tensor gives no mathematical constraints on the solution, however it is most prone to data errors, especially in case of poor sensors geometry in underground mines. The full tensor can be decomposed into volume change, linear compression/dilatation and shear motion. The trace null tensor is constrained by no net volume change at source. The third tensor is a pure DC, assuming the shearing motion at the focus only (Wiejacz 1991).

Although all types of solutions were calculated for every studied event, the quality of the solution (Fig. 2.7) differed due to numerical errors of solution and compatibility of direction of first arrival signs of the P wave. In case of events that

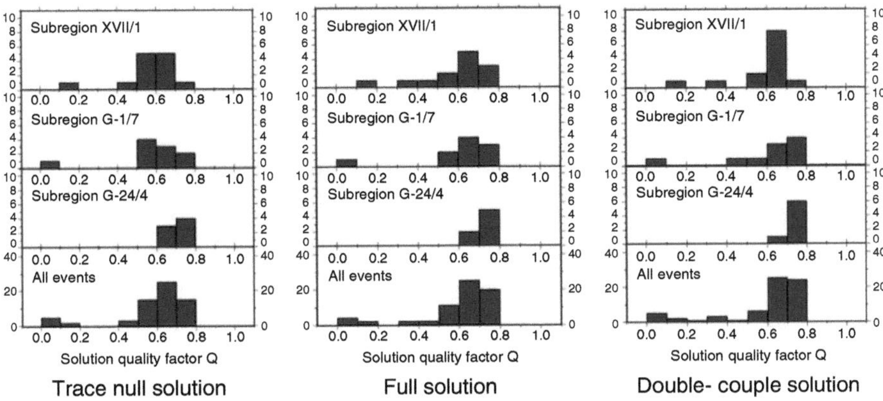

Fig. 2.7 Histograms of solution quality for trace null, full and DC solutions

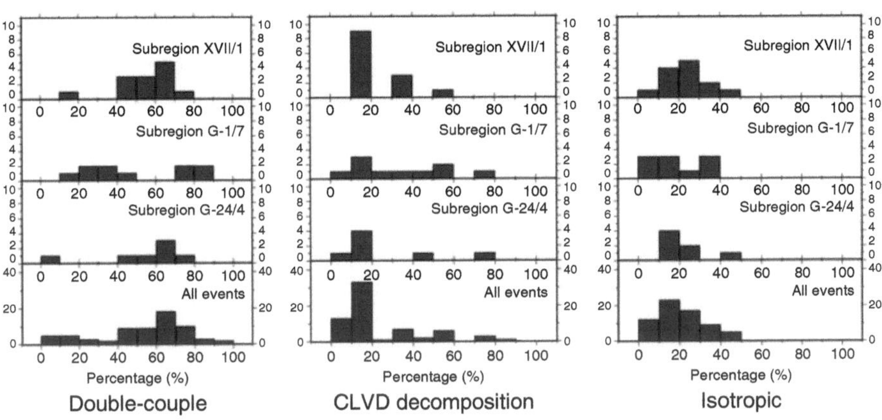

Fig. 2.8 Histograms of decomposition components for full tensor moment solutions

had taken place near borders of the local seismic network, the sensor distribution could have a negative influence on the quality of inversion of moment tensor. This is shown in Fig. 2.7 where quality factors are mostly between 0.6 and 0.8 with only a few events of poor quality (below 0.5).

Most of events had predominant DC component. A significant number of events had also some CLVD and isotropic components involved (Fig. 2.8). We did not observe any dominant fault plane for all events as well as for subregions analyzed separately.

The final results are presented for L1 norm solution. Although the Q factors for L2 norm often are higher, L1 norm solutions appeared to be more coherent for the studied events. Quality of solutions obtained with the use of norm L2 had a higher amplitude between the best and the worst cases than in the case of Q values for L1 norm (Fig. 2.7).

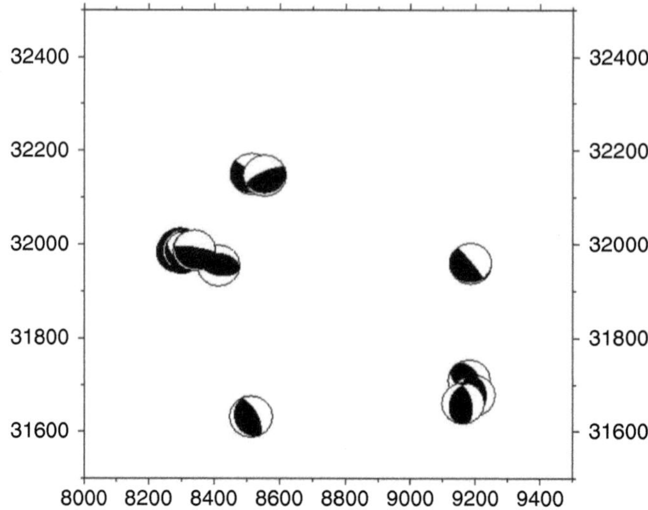

Fig. 2.9 DC solutions of moment tensor inversion for XVII/1 mining subregion of Rudna Copper Mine

2.5 Summary

We have analyzed 50 events from Rudna Copper Mine induced in the second half of 2007 obtaining spectral and physical parameters and focal mechanisms for every event. Radii of foci indicating sizes of focal areas were between 135 and 488 m. The dominant value was about 200 m. Thus, a typical event could have caused about 200 m of weakened area in rock mass around focal point where rupture processes had taken place.

The solution quality analysis of focal mechanisms calculated using full moment tensor inversion technique made us to choose L1 norm for presenting final results of mechanism solutions. This norm is characterized by lower sensitivity to data errors.

An analysis of fault planes orientations in the whole mine area as well as in its subregions does not lead to conclusions on any dominant fault plane orientation. It is unlikely that rupture processes of studied events are connected with any new activity on existing fault zones nor any known direction of cracks in the rockmass. The lack of dominant fault plane orientation may indicate that the studied events are mainly effects of post blasting relaxation and excavation of ores from beds. The studied period is too short to ensure us that the mentioned geological processes will not be initiated in the considered area of Rudna Copper Mine. It does, however, evidence that exploitation techniques had no detectable effect on tectonics.

In our opinion, a similar routine as described in our work, if periodically repeated, could provide additional information for mine engineers and can be used together with seismic event location and magnitude determination routines. The information about size and mechanisms of induced events can help in designing excavation works and blasting procedures in practice of stress release.

References

Aki, K. and Richards, P. G. (1980) Quantitative Seismology. Theory and Methods, Freeman, San Francisco.

Andrews, D. J. (1986) Objective determination of source parameters and similarity of earthquakes of different size. Earthquake Source Mechanics (S. Das, J. Boatwright, and C. H. Scholtz, eds.) Vol. 6, pp 259-267, Am. Geophys. Union, Washington, D.C.

Hanks, T. C., and Kanamori, H. (1979) A moment magnitude scale. J. Geophys. Res. 84, 2348-2350.

Madariaga, R. (1976) Dynamics of an expanding circular fault. Bull. Seism. Soc. Am. 66, 639-666.

Snoke, J. A. (1987) Stable determination of (Brune) stress drops. Bull. Seism. Soc. Am. 64, 1295-1317.

Wiejacz, P. (1991) Investigation of focal mechanisms of mine tremors by moment tensor inversion, Ph. D. thesis, Inst. Geophys., Pol. Acad. Sci., Warsaw (in Polish).

Chapter 3
Seismic Loading of Medieval Jeroným Mine During West Bohemia Swarm in 2008

Zdeněk Kaláb and Markéta Lednická

Abstract Seismic loading of underground spaces of Jeroným mine in 2008 was very low until 6 October 2008 when significant seismic loading occurred because of a very intensive seismic swarm that started in West Bohemia (Nový Kostel area near Kraslice). The source area was about 25 km away from the mine. Seismic waves from 451 earthquakes from Nový Kostel area were recorded in seismic station named JER1 within 6 October to 10 December 2008. The most intensive shock occurred on 14 October (21:00); maximum value of component velocity reached 0.435 mm s^{-1}. Fortunately, the earthquakes did not evoke visible damages in underground spaces.

3.1 Introduction

The Jeroným mine is a National Cultural Heritage Site of the Czech Republic located in Čistá in the Sokolov Region. At present, the whole system of mine workings (cavities) is divided into two separate parts, one called the Old Mine Workings (denoted OMW below) and the other the Abandoned Mine Workings (denoted AMW below). The Jeroným mine site forms a part of the European Mining Heritage Network. The OMW are specifically related to the extraction and processing of tin during the second half of the sixteenth century, whereas the AMW were created during the period of more than 400 years of exploration and extraction that took place afterwards (Žůrek and Kořínek 2004; Kaláb et al. 2006, 2008a; Žůrek et al. 2008).

Z. Kaláb (✉) • M. Lednická
Institute of Geonics Academy of Sciences of the Czech Republic, v.v.i., Studentská 1768, 708 00 Ostrava-Poruba, Czech Republic
and
VŠB-Technical University of Ostrava, 17. Listopadu 15, 708 33 Ostrava-Poruba, Czech Republic
e-mail: kalab@ugn.cas.cz; lednicka@ugn.cas.cz

A.F. Idziak and R. Dubiel (eds.), *Geophysics in Mining and Environmental Protection*, Geoplanet: Earth and Planetary Sciences 2, DOI 10.1007/978-3-642-19097-1_3, © Springer-Verlag Berlin Heidelberg 2011

The acquisition of information for predicting the stability of underground spaces is the main aim of geotechnical monitoring and iterative geophysical measurements in historical mines. This data is evaluated in relation to knowledge about the current state of underground spaces, especially fracturing and degradation of the rock mass. Results obtained with mathematical modeling of selected parts or of the whole complex of workings form the foundation of expert opinions on stability. Information obtained by geotechnical monitoring is essential for the creation of a reliable mathematical model. One site at which geotechnical monitoring has been implemented for this purpose is also the medieval Jeroným mine (Hrubešová et al. 2007) where measurements in the Abandoned Mine Workings are the basis for the case study briefly described here. This paper deals with seismic loading of mine workings especially during swarm in West Bohemia, Nový Kostel area, in 2008.

3.2 Geotechnical Observations

Quarterly geotechnical observations of the stability of selected parts of the AMW have been carried out since the year 2002. The convergence of mine workings along standard profiles, the fluctuations in the level of water standing in the mine and the growth of natural geological discontinuities in the rock mass have all been monitored (e.g. Žůrek et al. 2008; Kaláb et al. 2006). However, these measurements do not provide a record of the sudden fluctuations in these parameters that arise during short-term seismic events.

Since 2004, the seismic loads on the mine workings have been monitored, especially when blasting was being undertaken during the restoration of the drainage adit (Kaláb et al. 2008b). To obtain more objective and specific information about the state of stress-strain and the stability of this shallow mine, a program of experimental work was designed and, in 2005, a distributed measurement network (DMN) was built (Knejzlík and Rambouský 2008; Kaláb et al. 2008c). This network system is integrated with the existing seismic recording station transmitting data to the recording station in Ostrava via the GSM network. The monitoring system has a completely modular design so that the configuration of the system can be changed as required. At present, the distributed measurement network (DMN) has been tested using six different types of sensors (about 15 sensors are used now); however, it is possible to append up to a total of 250 sensors.

Communication within the distributed measurement network is controlled by computer. The DMN consists of a control unit, a communication bus (two-way communication) and remote addressed functional units. Communication between the control unit and the functional units takes place using a defined protocol. Concerning the reliability of operation of devices located over an extensive area, it is important that the functional units separate the input and output circuits from the Bus galvanically. Thus, the danger that disturbing foreign potentials (stray currents and lightning strikes) can disturb the system is eliminated. A recording device PCM3-MU (Knejzlík and Kaláb 2002) is installed in the chamber K1 in a

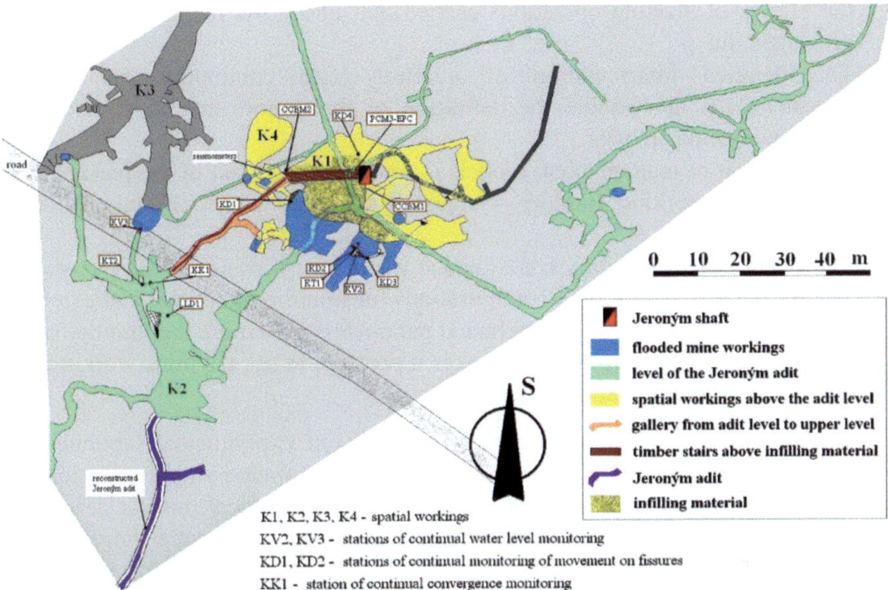

Fig. 3.1   Sketch of the Jeroným Mine and sensors of the DMN (for details see Kaláb et al. 2008c)

box with a cover IP65. The DCF time signal receiver, an insulator block and a GSM modem are installed in the hoist shaft near the surface. The Yagi antenna of the GSM modem is installed on a mast near the shaft.

At present, the DMN has the capacity to make the following measurements (Fig. 3.1):

- Continuous measurement of fluctuations in the level of the mine water table (2 measuring points).
- Continuous measurement of the opening (closing) of natural fractures in the rock mass (5 points).
- Measurement of the vertical direction of convergence (1 point).
- Measurements of the temperature of the mine atmosphere (2 points).
- Measurements of changes in the tensor of the stress state in the rock mass (2 points).
- Measurements of the floor to ceiling height of the stope using a laser distance measuring device (1 point).
- Measurements of the seismic load on the mine working (three-component recording).

The installation of the DMN at Jeroným mine has led to the following scientific benefits:

- The assessment of the load on the mine working caused by natural seismicity in the Kraslice region, and also the load due to induced seismicity, especially the effects of blasting operations carried out during the reconstruction of the drainage adit.

- The detailed interpretation of quarterly and continuous geotechnical measurements that enable the definition of the factors that most affect the stability of mine cavities.
- The creation of mathematical models (variant calculations) of the present state of the mine working stability – chamber K2, into which the reconstructed drainage adit runs.
- The definition of boundary conditions at which the workings become unstable.
- The proposal of operational configurations for minimum, optimal and extended continuous monitoring of the technical parameters governing the stability of the Jeroným mine (the current array makes it possible to use the mine working as a natural experimental laboratory).

The Jeroným mine is thus already in use as an experimental laboratory to evaluate new instrumentation for detecting rock deformation and seismic loading and the measurements are being used to construct theoretical models describing these phenomena. Some of the most interesting measurements are those relating fluctuation of mine water level at location KV2 (Kaláb et al. 2008d). In the lower levels of the workings there is a definite correlation between sudden decreases in mine water level and seismic loading of the mine workings. One example of the new instrumentation being tested is the compact conical strain gauge probe (Knejzlík et al. 2008) used to measure stress in borehole overcoring of the rock mass.

A most important conclusion based on the measurements obtained from the DMN is that the mine workings are stable. Thanks to this, it is possible to continue with preparations for opening of the Jeroným mine heritage site to the public.

3.3   Seismic Loading of Mine Workings

Detailed analysis of seismic loading of the Jeroným mine and its surroundings was prepared during designing of the restoration of the drainage adit in 2003 (Kaláb 2003). This study documented that mine workings can be influenced by more intensive earthquakes from the West Bohemia (Kraslice region) and that technical seismicity is very low here. Technical seismicity is represented by blasting operations from adjacent quarries and traffic from the road above the mine. However, technical seismicity will significantly increase during restoration of mine spaces. Particular measure is necessary to take into account if is used as part of technology for driving. In that case, induced vibrations can exceed acceptable (conventional) values and can start damaging this heritage. Information about seismic regime in Kraslice region during last decades can be found in the literature (e.g. Composite authors, 2000).

Original drainage adit was in many parts slumped and buried and therefore it did not meet its dewatering and ventilation needs. The new adit also provided another entrance to underground spaces of the AMW. After opening it, seismic monitoring was started using seismic station signed as JER1 (e.g., Kaláb and Knejzlík 2004; Kaláb et al. 2006). Particular aims of seismic monitoring were:

- To maximize breakup without failure of rock in the surroundings
- To minimize seismic effect in underground spaces
- To use the most value of boreholes with explosives and specific timing.

3.4 Seismic Swarm in West Bohemia

In total, 722 events were registered by seismic station JER1 in 2008 using triggered regime. Seismic loading of mine workings can be defined as low up to 6 October 2008. This situation changed after the occurrence of seismic swarm in West Bohemia, Kraslice region. Since 6 till 30 of October 2008 more than 20,000 events occurred at the southern edge of the Nový Kostel area. The hypocenters lie in the depth range 10.5–9 km (below the microswarm of February 2007) and have reactivated the southern patch of the 2000 swarm. Several $M > 3.5$ events have occurred making this swarm the strongest one since the $M = 4.6$ swarm of 1985/1986. The swarm is exceptional due to unusually fast release of energy ($>10,000$ events in 7 days). The most intensive earthquake occurred on 12 October, 09:44 (Central European Time Zone) with magnitude 3.9 and was felt in wide surroundings, for example also in western part of Prague. Following intensive events occurred on 14 October, 21:01, with $ML = 3.9$ and 28 October, 09:30, with $ML = 3.7$ (according www.ig.cas.cz). The end of this swarm is dated to 10 December 2008.

As it is documented by Geophysical Institute of the ASCR, Prague, the release of seismic energy was unbalanced, both concerning event intensity and chronology of occurrence. The Jeroným mine is located about 25–30 km south-east from the source area in Nový Kostel. The number of recorded events significantly increases and consequently, seismic loading of the mine working occurred. To have an adequate quantization of seismic signals of weak seismic events during the period with low seismic loading, the maximum recorded value was determined to 0.25 mm s^{-1}.

The JER1 station recorded altogether 451 earthquakes that occurred in the Nový Kostel area between 6 October and 10 December 2008. Figures 3.2 and 3.3 graphically present values of maximum recorded velocity of given events in time sequence. Figure 3.2 shows all recorded earthquakes and Fig. 3.3 displays only weak earthquakes with the peak velocity value up to 0.05 mm s^{-1}. The maximum velocity value was recorded at JER1 station during 14 October earthquake whereas component value of velocity was 0.435 mm s^{-1}. This wave pattern is presented in Fig. 3.4.

Acceptable seismic loading and possible damages of special structures are determined on the basis of expert opinions. This way was used also for determining limit velocity values induced during drainage adit restoration in 2003. Therefore, maximum recorded value of velocity was only 0.16 mm s^{-1} (recorded on 10 January 2006). Though, this maximum value was exceeded three times in

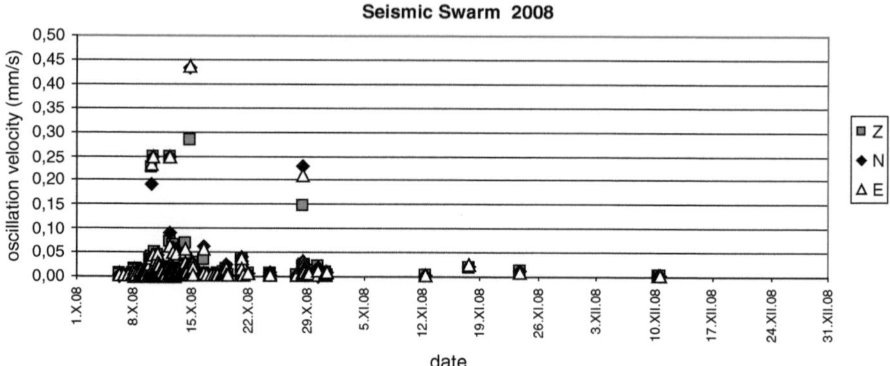

Fig. 3.2 Chronology of all earthquakes of the seismic swarm occurred near Nový Kostel in 2008 recorded at the JER1 seismic station (vertical: maximum oscillation velocities)

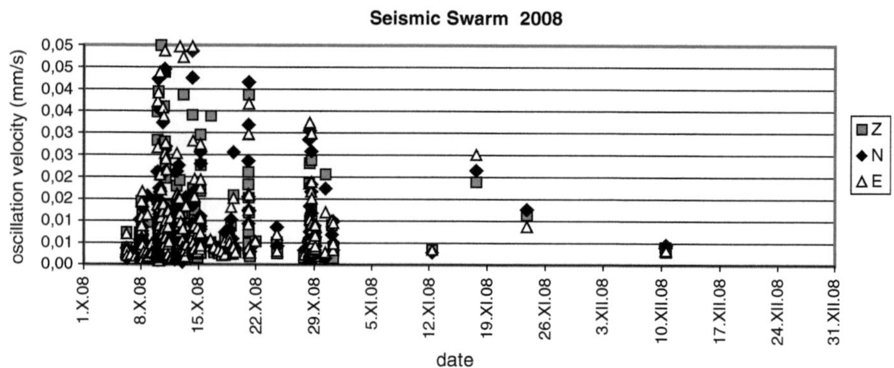

Fig. 3.3 Chronology of weaker earthquakes of the seismic swarm occurred near Nový Kostel in 2008 recorded at the JER1 seismic station (vertical: maximum oscillation velocities)

comparison with the maximum recorded velocity value from the strongest event (0.435 mm s^{-1}) that occurred during last swarm. Quick visual scan of mine workings, which was made during quarterly observing and geotechnical monitoring in autumn 2008 and spring 2009, did not document damages evoked by earthquakes from the swarm. It means that larger breaking off stones, opening of natural fractures in rock massif and/or new fissures in pillars were not found. Minor cracking of the rock walls and ceilings cannot be detected visually. Measurement of convergence using mechanical instrument did not give significant differences from earlier measurements. Results from the DMN are presented in this issue. It is possible to derive that we can accept higher limit velocity value induced by building operations – perhaps up to 0.4 mm s^{-1}. However, it is necessary to evaluate an

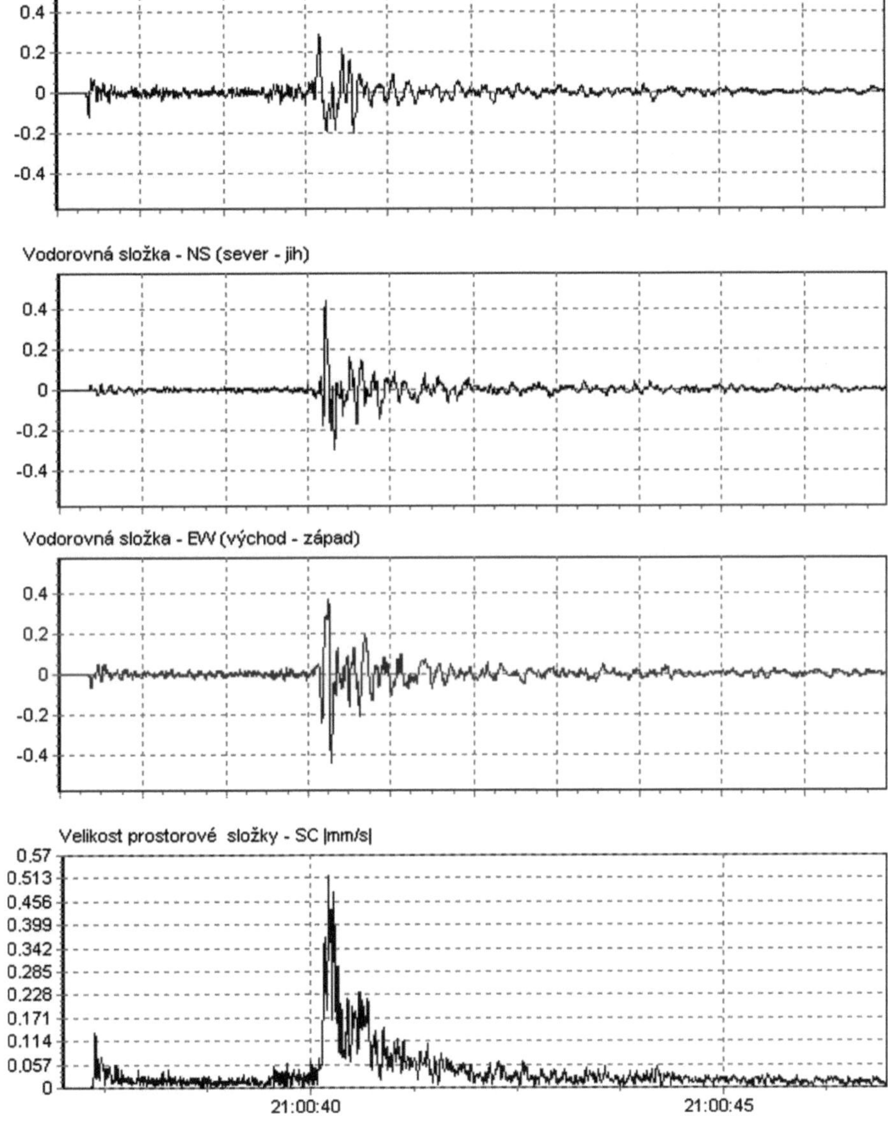

Fig. 3.4 Wave pattern of microearthquake from Nový Kostel region, 14 October 2008, 21:00, recorded at Jeroným station (top down: vertical, horizontal N–S, horizontal E–W and space components)

influence of different frequency ranges and durations of evoked technical vibrations. Spectral analysis of blasting operations from 2001 determined frequency range of records as 4–60 Hz (the source–seismic station distance of about 300 m). Nevertheless, the frequency range of microearthquake records is 2–25 Hz only.

3.5 Conclusion

This paper summarizes information about seismic loading on medieval Jeroným mine during swarm in West Bohemia, Nový Kostel area, in 2008. The source area of Nový Kostel swarm is about 25–30 km far form the mine workings. The Kraslice region and surroundings, in which Nový Kostel is located, were rejuvenated in the Tertiary and tectonic movements and remains seismically active up to now. Using the map of seismic risk of the Czech Republic from Supplement of Eurocode 8, earthquakes with macroseismic intensity up to 6.5° can be predicted in the area under discussion.

The JER1 seismic station that is located in mine workings of medieval mine under discussion recorded more than 450 events during the swarm in 2008. The maximum velocity value was recorded at JER1 station during 14 October earthquake whereas component value of velocity was 0.435 mm s^{-1}. Damages of underground spaces (i.e., cracking of pillars, opening of observed fissures and discontinuities, more significant breaking off rock from the ceilings and walls …) were not observed during quarterly experimental geomechanical monitoring in autumn 2008 and spring 2009.

The main aim of the described seismological monitoring is to obtain information about seismic loading of historical mine workings – National Cultural Heritage Site of the Czech Republic, Jeroným Mine. However, recorded data can be used for scientific purposes.

Acknowledgement This paper has been achieved with the financial support of the Czech Science Foundation, project No. 105/09/0089.

References

Composite authors: Special Issues of Studia Geophysica et Geodaethica, Vol. 44/2 and 44/4, AS CR, Prague

Hrubešová, E., Kaláb, Z., Kořínek, R. and Žůrek, P. (2007) Geotechnical Monitoring and Mathematical Modelling in Medieval Mine Jeroným (Czech Republic). Górnictwo i Geoinżynieria, Vol. 31, Zeszyt 3, 183–190.

Kaláb, Z. (2003) Evaluation of Seismic Loading of Medieval Jeroným Mine in Czech Republic. Acta Montanistica Slovaca. Ser. 8(2003), No.1, Košice, Slovakia, 36-41 (in Czech).

Kaláb, Z. and Knejzlík, J. (2004) Experimental Measurement of seismic Effect of Blastings in Historical Mine Jeroným. Transactions, Vol. IV, No. 2/2004, 159–166 (in Czech)

Kaláb, Z., Knejzlík, J., Kořínek, R. and Žůrek, P. (2006) Cultural Monument Jeroným Mine, Czech Republic – Contribution to the Geomechanical Stability Assessment. Publs. Inst. Geophys. Pol. Acad. Sc., M-29(395), 137–145.

Kaláb, Z., Lednická, M. and Hrubešová, E. (2008a) Mining History and Present State of Medieval Mine Jeroným. Zeszyty Naukowe Politechniki Śląskiej, Ser. Górnictwo Ser. 283, Nr. 1781, Gliwice, 61–70.

Kaláb, Z., Knejzlík, J. and Lednická, M. (2008b) Seismic Station with Geomechanical Network in Medieval Mine. Proceedings of European Seismological Commission ESC2008, 31st General Assembly, CD - Short Papers electronic volume, 167–174.

Kaláb, Z., Knejzlík, J., Kořínek, R., Kukutsch, R., Lednická, M. and Žůrek, P. (2008c) Contribution to Experimental Geomechanical and Seismological Measurements in the Jeroným Mine. Acta Geodynamica et Geomaterialia, Vol. 5, No. 2(150), 213–223.

Kaláb, Z., Hrubešová, E., Knejzlík, J., Kořínek, R., Kukutsch, R., Lednická, M. and Žůrek, P. (2008d) Mine Water Movement in Shallow Medieval Mine Jeroným (Czech Republic). In: Rapantová, N. and Hrkal, Z. (Eds): Mine Water and the Environment. Proceedings of 10th International Mine Water Association Congress. Karlovy Vary. VŠB-Technical University of Ostrava, 19-22; full paper at CD, No. 37, 11 pages.

Knejzlík, J. and Kaláb, Z. (2002) Seismic Recording Apparatus PCM3-EPC., Publs. Inst. Geophys. Pol. Acad. Sc., M-24(340), 187–194.

Knejzlík, J. and Rambouský, Z. (2008) Current Solution for Distributed Control and Measurement System in the Jeroným Mine – Modular System. Acta Geodynamica et Geomaterialia, Vol. 5, No. 2(150), 205–212.

Knejzlík, J., Rambouský, Z., Souček, K. and Staš, L. (2008) Second Generation of Conical Strain Gauge Probe for Stress Measurement in Rock Massif, Acta Geodynamica et Geomaterialia, Vol. 5, No. 3(151), 257–265.

Žůrek, P. and Kořínek, R. (2004) Opening of the Medieval Jeroným Mine in the Czech Republic to the Public, Acta Montanistica Slovaca, Ser. 8., 96–100.

Žůrek, P., Kořínek, R., Kaláb, Z., Hrubešová, E., Knejzlík, J., Daněk, T., Kukutsch, R., Michalík, P., Lednická, M. and Rambouský, Z. (2008) Historical Mine Jeroným in Čistá. Monograph, VŠB – Technical University of Ostrava and Institute of Geonics ASCR Ostrava, 82 pages (in Czech).

Chapter 4
Spin and Twist Wave Groups in the Seismic Field: A Case of Mining Event

Krzysztof P. Teisseyre

Abstract In the theory of rupture and rupture preparation in a medium treated as asymmetric continuum, important role is provided for rotational motions and asymmetric stresses. The rotational motions generate oscillations which propagate in the form of rotational waves, provided that the elastic properties of the medium are asymmetric or the medium consists of loose grains.

Therefore, we found it very important to identify the traces of rotational motions which have occurred in the seismic source. Such traces should be present in the recording of the seismic event, and in the seismic noise just preceding it. This work presents results of such a quest in the seismic field generated by an event in Polish Upper Silesian coal mine, and detected at two distant observatories.

The rotational components of the seismic field, spin and twist, were detected with rotational seismometers. Then, distinct groups of waves – or pulse sequences – were sought in the curves of both components. Pulses of one component were compared to the other directly and after transformation (phase-shifting and sign reversal), because in the source, the relations between spin and twist motions should include time-lapses which translate into phase-shifting of the generated oscillations.

Several time-periods of inter-component conformance just before and during the seismic event were found; these include some conformance episodes between rotational oscillations found at different observatories.

4.1 Introduction

Our considerations are based on the theory of correlated processes in a seismic source. We follow the asymmetric continuum theory, in which equations of motion and constitutive laws include the fields of both displacement and rotation. In a

K.P. Teisseyre (✉)
Institute of Geophysics, Polish Academy of Sciences, Księcia Janusza 64, 01-452 Warszawa, Poland
e-mail: kt@igf.edu.pl

A.F. Idziak and R. Dubiel (eds.), *Geophysics in Mining and Environmental Protection,*
Geoplanet: Earth and Planetary Sciences 2, DOI 10.1007/978-3-642-19097-1_4,
© Springer-Verlag Berlin Heidelberg 2011

source region, of importance are changes of strain and release processes; the displacement equation of motion can be transformed to the wave equations (Teisseyre 2009). Preseismic accumulation of local deformations leads to a local stress increase, followed by release in a rebound process, which can be shifted in phase. Thus, we consider the pulsing deviatoric strain, which is the oscillation of local shear axes and the shear amplitude. Such a field is called further on a shear-twist or shortly *twist*; moreover, we consider the rotation field as a second source of rotational oscillations, called the rotation or *spin* waves. The coincidence of shear-twist and spin in the fracturing processes was described by Teisseyre et al. (2008). For description of basic deformations and motions, see Teisseyre and Górski (2008, 2009).

Here we insert a simple explanation of what the spin and twist are. These motions are three-dimensional oscillations. Let us look at the vertical rotation (that is, around the z axis) because in this paper we deal only with this, horizontal aspect of motions. In a rotating perfect rigid body, the spin or rotation in the x, y plane is

$$\omega_z = \frac{\partial \dot{u}_y}{\partial x} = -\frac{\partial \dot{u}_x}{\partial y} \tag{4.1}$$

where $\dot{u}_i$ stands for velocity of displacement in the i direction. The spin does not depend on the position of horizontal axes, as any pair of axes in plane x, y may be the main axes.

If the body is prone to deformations, these spatial derivatives may vary in time. For any given moment and area, rotation/spin is the arithmetical mean of two spatial derivatives:

$$\omega_z = \frac{1}{2}\left(\frac{\partial \dot{u}_y}{\partial x} - \frac{\partial \dot{u}_x}{\partial y}\right) \tag{4.2}$$

The measure of twist, which may be called the rotation anisotropy, is

$$E_{xy} = \frac{1}{2}\left(\frac{\partial \dot{u}_y}{\partial x} + \frac{\partial \dot{u}_x}{\partial y}\right) \tag{4.3}$$

Signs "$-$" and "$+$" in (4.2) and (4.3), respectively, are a consequence of the coordinate system used.

Twist is part of the deformation tensor. The value and sign of this component depend on x and y axes position vs. directions of the sensors' motion. For spin there is no such relation, provided that sensors swing in the plane x, y.

In experimental practice, each spatial derivative is substituted by *differential signal* obtained from a pair of sensors. The differential signal or component is understood to mean the difference between two signals divided by the distance between sensors. The *initial* signals are here the velocities of linear displacements (the translational motions, as opposed to rotational). The initial signals are counted in m/s, while the differential and rotational signals have another unit: rad/s.

According to the theory, rotational components of the waves generated by a seismic event are related to the processes in the source – the fracturing and fracture preparation. Two kinds of these waves, twist and spin, may be inter-correlated: directly or with reversed sign, or with a phase shift of $\pi/2$. The phase shift means a retarded process. So, the correlation may be between spin and twist, twist and retarded spin, etc. Then, it is advisable to study rotational components in first sections of the recorded event, when seismic waves reflect, at least to a certain degree, processes occurring when the fractures start and proceed. Later, various delayed and secondary waves superimpose, obscuring the picture which already was complicated.

Besides, episodes of correlated, i.e., synchronized rotational oscillations may be masked by other motions. This pertains especially to the retarded processes, because time-shift due to the phase-shift depends on the signal frequency. Therefore, we are searching for twist-spin coincidences, direct, reverse in sign, and with phase shift, after applying band-pass filtration to the data.

The purpose of band-pass filtration is to facilitate seeking the same waveforms (at least few wavelength in duration), contemporaneous in the compared signals. To obtain the above-mentioned phase shift, we apply the Hilbert transform to one of the signals, twist or spin.

If we assume that processes in the source, such as deformation, fracturing and slipping, are accompanied by rotational motions, then we may expect similar motions before the major fracturing – in the preparatory stage. Of course, these motions should be small and the waves generated may be indistinguishable from the background noise. Nevertheless, we pay attention to any prominent or synchronized wave groups visible in some band-pass filtered, rotational signals just before the onset of the waves generated by the proper seismic event. Especially, cases of twist-spin synchronization with phase leap (by one of components) of $\pi/2$ are promising, as it is unlikely that such coherences originate accidentally, not in the dynamic processes in the focus.

The research on seismic rotational oscillations and their sources done at the Institute of Geophysics is both theoretical and experimental. The theoretical approach is focussed on the development of theory of rotational motions in the seismic focus together with theories explaining how the ensuing rotational waves should propagate. The spin and twist motions appear to be effectual occurrences in the seismic focus, during a rupture and before, in the stage of preparation. These motions generate rotational waves. The theory of propagation of these waves develops in the frame of the asymmetric continuum theory. Asymmetry of stresses is treated as a necessary feature which enables the propagation of rotational waves (Teisseyre et al. 2006, 2008).

The experiments performed up to now were based mainly on detecting rotational and other differential signals, in various natural environments; in regular seismic stations (Teisseyre et al. 2003a) and also on/near the glaciers (Teisseyre et al. 2004; Górski and Teisseyre 2006), in the course of field surveys.

Relations between classic and rotational records were considered in quite a few of our papers, with attempts to estimate the parts of energy radiated in the form of twist and spin waves (see Teisseyre et al. 2003a). The present article originates from our willingness to check whether spin-twist relationship in the focus, as anticipated by the theory, could be confirmed by the recorded seismic data, more exactly – in the waveforms. We were especially interested in the first parts of seismograms which are not disturbed by delayed waves.

4.2   Methods

In this work, an analysis of only one seismic event is presented. Its source was detected by two Polish observatories: it was M $=$ 3.1 mining event which occurred on 8 July 2005 at 21.10:55.92, in Upper Silesian Coal Basin, about 180 km SE from Książ and about 76 km NW from Ojców. Good simultaneous records of rotational seismometers in Książ and Ojców are rare – although both rotational stations operate for several years, they not always work in the same time.

At both stations, Książ and Ojców, identical microarrays of two rotational seismometers operate. At each site, the seismometers are positioned close and perpendicular to each other. Our rotational seismometer is a pair of pendulum seismometers (SM-3 type), mounted in horizontal mode in one chassis in such a way that they have a common, vertical axis of swing. One seismometer is mounted on top of the other, in the upside-down position. This apparatus has been already described by Teisseyre et al. (2003b). From one rotational seismometer at the recording station, we obtain one pair of digital signals – either channels 1 and 2, or 3 and 4. These initial signals are registered with the sampling rate of 100/s. From each pair, we obtain the differential signal. Further, from two differential signals we obtain the couple of rotational signals: spin and twist. However, we do not get the exact values of twist, because of uncertainty of its axes, and because these axes oscillate too (slightly). To get the exact twist amplitudes, and localization of the twist axes, a more complex seismometer microarray is needed.

We looked through the original seismograms. Then, chosen time-periods were analyzed, using procedures in Matlab® (also the filtration procedures were developed and operated in the Matlab® environment). An important topic must be touched which relates to the technique: beside the Hilbert-transform shift mentioned in the Introduction, another phase shifting is introduced as an artefact of the filtrating procedure. This *filtration shifting* translates to time-shifting, and to a certain small distortion of the wave shape. Yet we are convinced that such a shortcoming does not preclude the usefulness of the procedure here. We have obtained all the signals of a given frequency band with the same procedure, so the time-shift should be similar. In the analysed cases, the obtained narrow-band signals are delayed relative to original signals. Estimation of this time-shift was possible by comparison with results of another, symmetric filtration procedure. For middle frequencies, from 3 to 12.5 Hz, this delay was about 0.5–0.7 s. Therefore, twist may be compared to spin, and comparisons of signals from different stations are also allowed. Besides, all the filters we used were of Chebyshev II-type, and have been attuned with the aim not to infer visible distortions. The above-mentioned symmetric filtration allowed for an additional test: certain seismic signals obtained when pass-band was 7.5–12.5 Hz were practically the same as those obtained by summation of two signals of more narrow spectra: 7.5–10 and 10–12.5 Hz. To recapitulate, the unwanted effects of filtration do exist, but they will harm mainly the comparisons between signals of various frequency bands obtained from the same recordings.

The chosen part of an original signal goes through the procedure which divides it into frequency bands; this is done for all four channels. Subsequently, drawing of the rotational components is performed for each frequency band.

After looking through results of this data processing, the examined case becomes subject to the next stage, provided that the noise level is not too high. The purpose of this new stage is to allow visualisation of the spin–twist relations. The spin and twist signals, of given frequency band, are loaded into the procedure which produces four new signals: twist shifted by $\pi/2$ (through the Hilbert transform, as in other instances); twist with inverted sign; twist shifted by $\pi/2$ and then inverted by sign; and finally – spin shifted by $\pi/2$.

Then we scrutinise the obtained diagrams, searching in each horizon for any stretches of conformity in the pair of signals, at least about ten samples long (0.1 s) or longer, where "conformity" means rough alignment of signals compared in a given horizon, or at least certain conformity of the shape only – in a chosen pair one signal may be bigger or smaller than in the second signal. A diagram obtained in described way was, however, illegible in this aspect, as it contained too long signals – at least several thousand samples. Therefore, it was necessary to make several descendant diagrams, each enlacing reasonable time-span, for example 4 s (400 samples). We paid attention mainly to the time-spans just before the onset of distinct waves which came from the shock focus (here: P-waves), and these embracing a short period after arrival of first waves and before the next phase arrives.

To make the image more clear, we label the episodes of wave group conformance as follows:

"+" means the conformance of unchanged spin and twist episodes
"−" means the conformance between the episodes of spin and twist reversed by sign
"H twist" and "Ht" stand for twist shifted by $\pi/2$, accompanied by spin
"−Ht" denotes a pair of spin and reversed shifted twist signal
"H spin" and "Hs" denote a pair of twist and shifted spin signal
"Hs − t" is for a pair consisting of reversed twist and shifted spin signal.

In most diagrams, these portions of curves to which no pair conformance can be ascribed were erased, leaving blank background. Therefore, it was possible to gather different conformity cases ("+", "−", etc.) in the same diagram. Those episodes of spin-twist agreement which look as transient casual coincidences, were not included into these diagrams. For comparison, the accompanying upper diagrams, in Figs. 4.2 and 4.4, show intact spin and twist curves.

Episodes of "H twist"-type are always similar to "Hs, −t", and the "−H twist" episodes are similar to "H spin". Therefore, it is sometimes not easy to decide which candidate for paired time-stretch is most probably the real one, as two almost perfect pairs coincide.

Showing both the twist and the reversed twist has two reasons; first – the theory describing twist-spin coupling permits generation of the related component with reversed phase; second – the sign of detected twist motions depends on twist axes position in relation to the detecting apparatus. We must state that in many cases we could not find episodes of any spin-twist conformity, beside short, apparently casual moments in higher frequencies which are hard to interpret.

4.3 Gross Results

At first glance, the results give grounds for making three observations:

1. The general course of initial and rotational signals is roughly congruent.
2. Usually the most clear pictures of the seismic event appear in the middle frequency bands, namely 7.5–10 and 10–12.5 Hz.
3. If original signals are afflicted with high noise, calculation of rotational signals produces bad pictures too.

The biggest amplitudes are found either in the above-mentioned frequency bands, or in lower frequencies. In most cases, analysis in higher frequencies is hampered by sharp peaks, especially numerous in the curves of twist; clearly, these peaks are caused by disturbances. Such a noise abounds also in other cases, not presented here. In middle frequencies, curves of spin and twist coincide in the record of the seismic event quite often. Theoretically, this could be the effect of notorious dominance of one differential signal over the other; but we checked it and it was not the case.

This approximate gross conformity between linear and rotational seismograms is not a new result. For example, Moriya and Teisseyre (2006) published diagrams that showed differential components of the seismic field produced by earthquakes which occurred at various distances. Both components obtained from pairs of horizontal seismometers, namely the NS component (transverse to NS axis) and EW component, were generally in conformity with translational components. Plots of rotation/tilt components (named "UD" by those authors) appeared more variable, but the earthquakes left their scores also in the curves of these components.

Cochard et al. (2006) show a general similarity between the record of rotation waves and the classic, translational seismograms. They also found that recorded spin motions agree with these calculated from an array of seismometers.

At first glance, all these conformities support the thesis of the focal provenience of rotational waves. Other explanations should be mentioned, however. The rotational motions may be generated by the medium's response to the passing seismic wave, especially at seismic boundary. Further, the rotational signals may be seriously contaminated by the influence of translational, "classic" seismic signals (Wiszniowski and Teisseyre 2010).

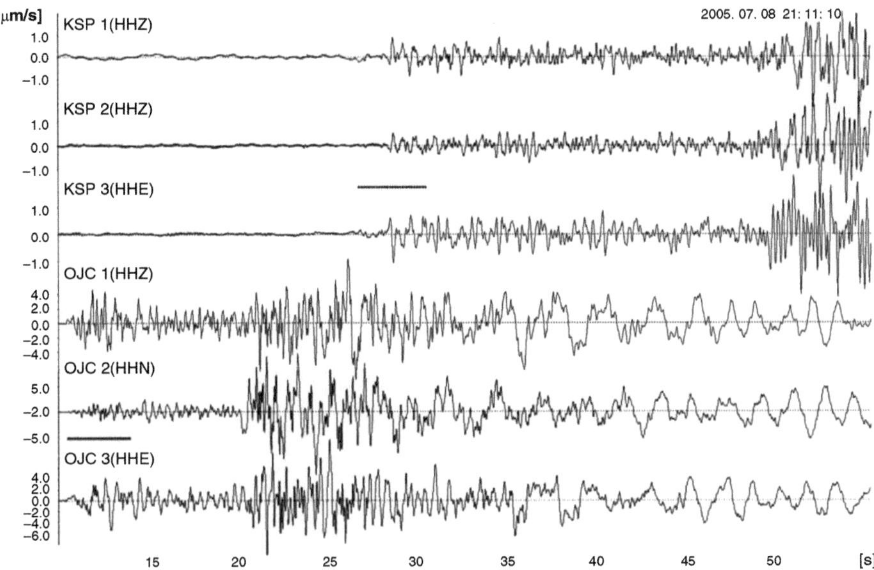

Fig. 4.1 Seismograms of the mining event in Upper Silesia, Poland, 2005.07.08, 21:10:55.92, M = 3.1, as in all figures. Records from regular seismic stations; KSP – Książ, components from *top* to *middle*: Z, N and E; OJC – Ojców, from *middle* to the *bottom*: Z, N, E. *Black horizontal lines* between the curves of horizontal components – the time-span from the apparent arrival of *P*-waves, approximately to the end of the signal windows depicted in Figs. 4.2–4.5; it is 3.5 s for Książ and 3 s for Ojców. *P–S* delay is 22.8 s for Książ and 10.7 s for Ojców. Horizontal scale – the UTC time of record

Recent investments in Książ Observatory made it possible to prove that rotational motions recorded there are not mere instrumental artefacts. First, operation of rotational seismometers microarray has been restored. Then, a new optical sensor of rotation (*spin*) has been installed there and its records of spin, from a few mining quakes of magnitudes ≥3.0, are conformant with those obtained with the microarray. This autonomic sensor has been constructed in the Military University of Technology in Warsaw (Jaroszewicz et al. 2010).

Figure 4.1 shows translational seismograms from the analysed seismic event recorded at regular seismic stations in Ojców (OJC) and Książ (KSP); these data have been obtained from the broadband STS-2 seismometers.

4.4 The Inter-Related Spin and Twist Curves

We chose for the analysis the low and middle frequency bands, up to 10 Hz. In higher frequencies, the analysis is more laborious and difficult, as the short congruent waveforms are quite numerous and it is hard to find any order in their distribution. It is often a matter of guessing, whether two groups are similar because of their homology, or just by chance. Besides, we have concentrated on the time-period from few tens of seconds before the start of the proper seismic event trace, to few seconds after this start.

For the case of the mining event, we note general rough conformance of the curves of spin and twist, up to 7.5–10 Hz frequency band. The results of analysis are presented in the form of diagrams for two frequency bands, 7.5–10 and 3–7.5 Hz. The effects of searching for congruent waves of the twist and spin components are shown in Figs. 4.2–4.5.

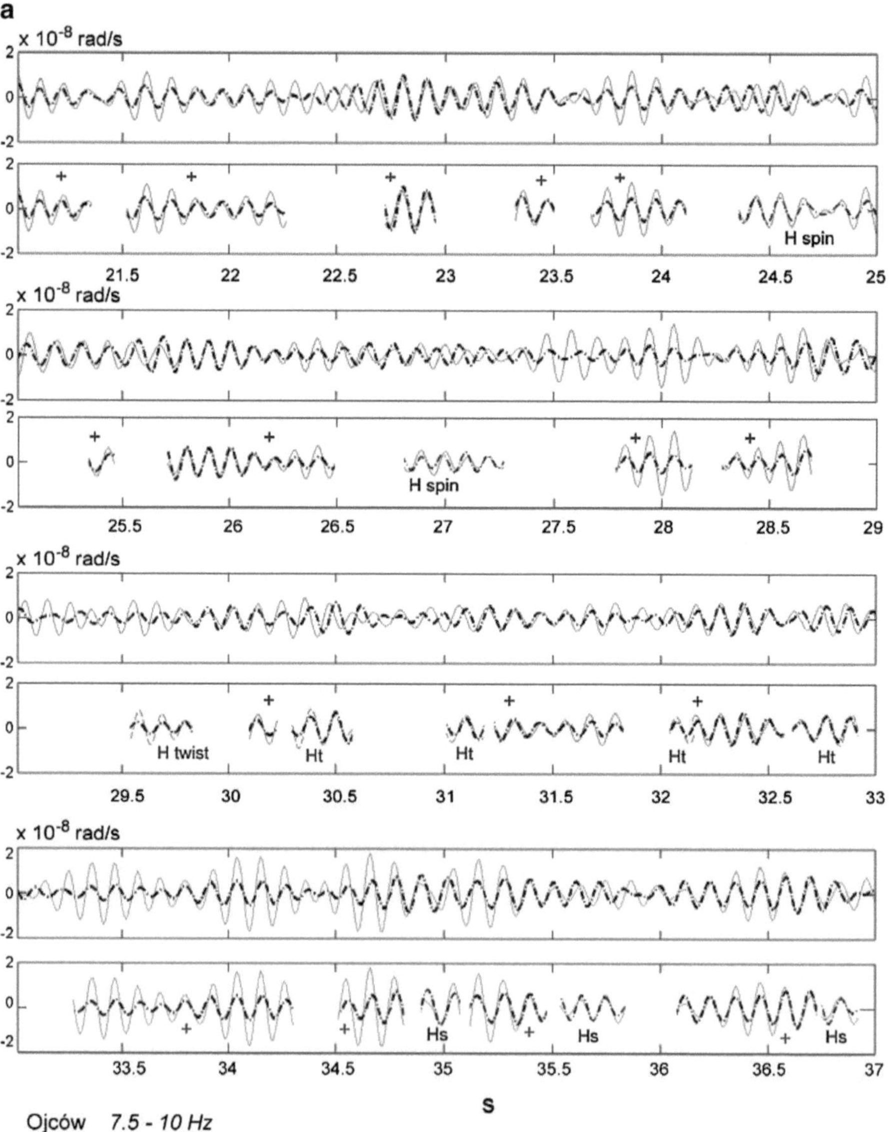

Fig. 4.2 continued

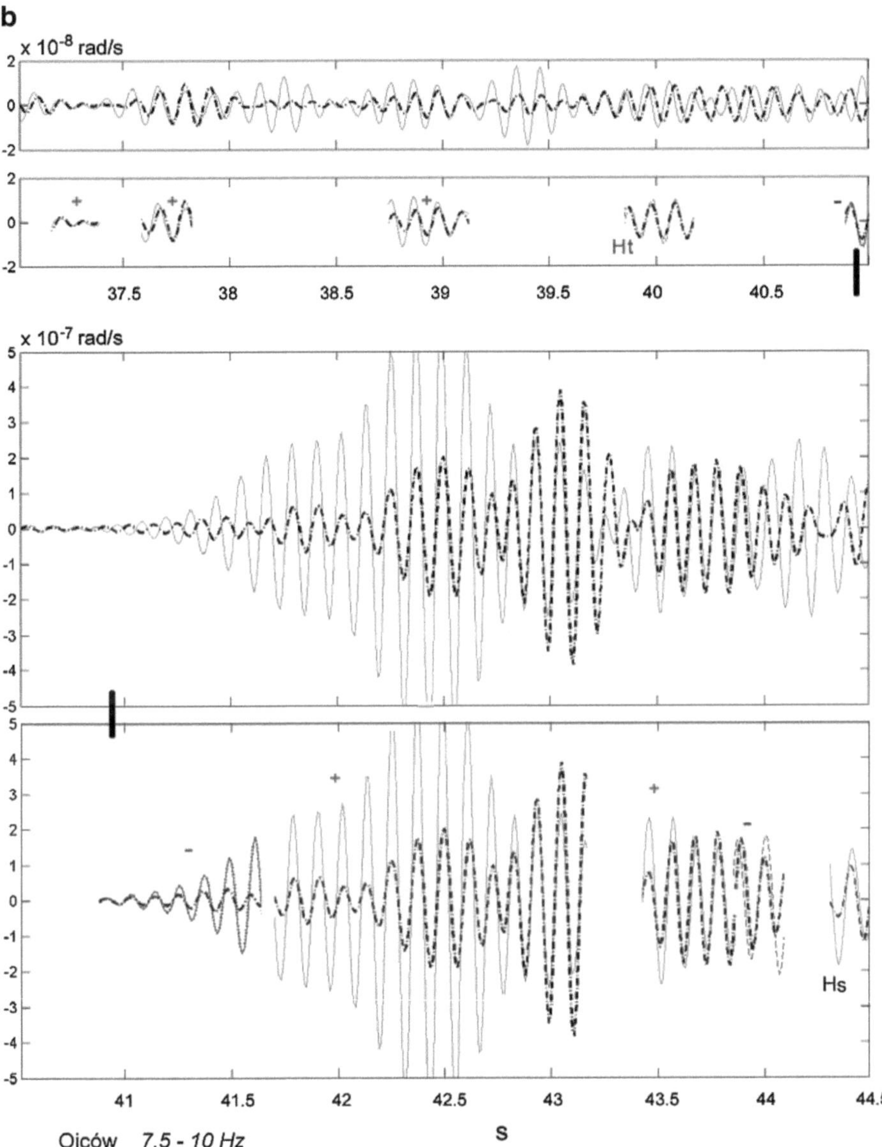

Fig. 4.2 Episodes of wave-groups conformity, 7.5–10 Hz. The seismic mining event, the same in all figures. The diagrams are paired, in each pair the *upper diagram* shows the spin and twist. Analysis of the recordings from rotational seismometers in Ojców; (**a**) first part; (**b**) second part including the onset of P waves. *Black vertical rod* – approximate position of apparent arrival of P-waves in these signals. *Symbols* "+" denote the conformance of unchanged spin and twist waveforms; "−" shows the conformity of reversed twist with spin (*dash-dotted*); "H twist", "Ht" show the conformance of shifted twist and spin ; "H s, −t" show the conformity between shifted spin and reversed twist. Horizontal scale: time after start of the time-period chosen for analysis

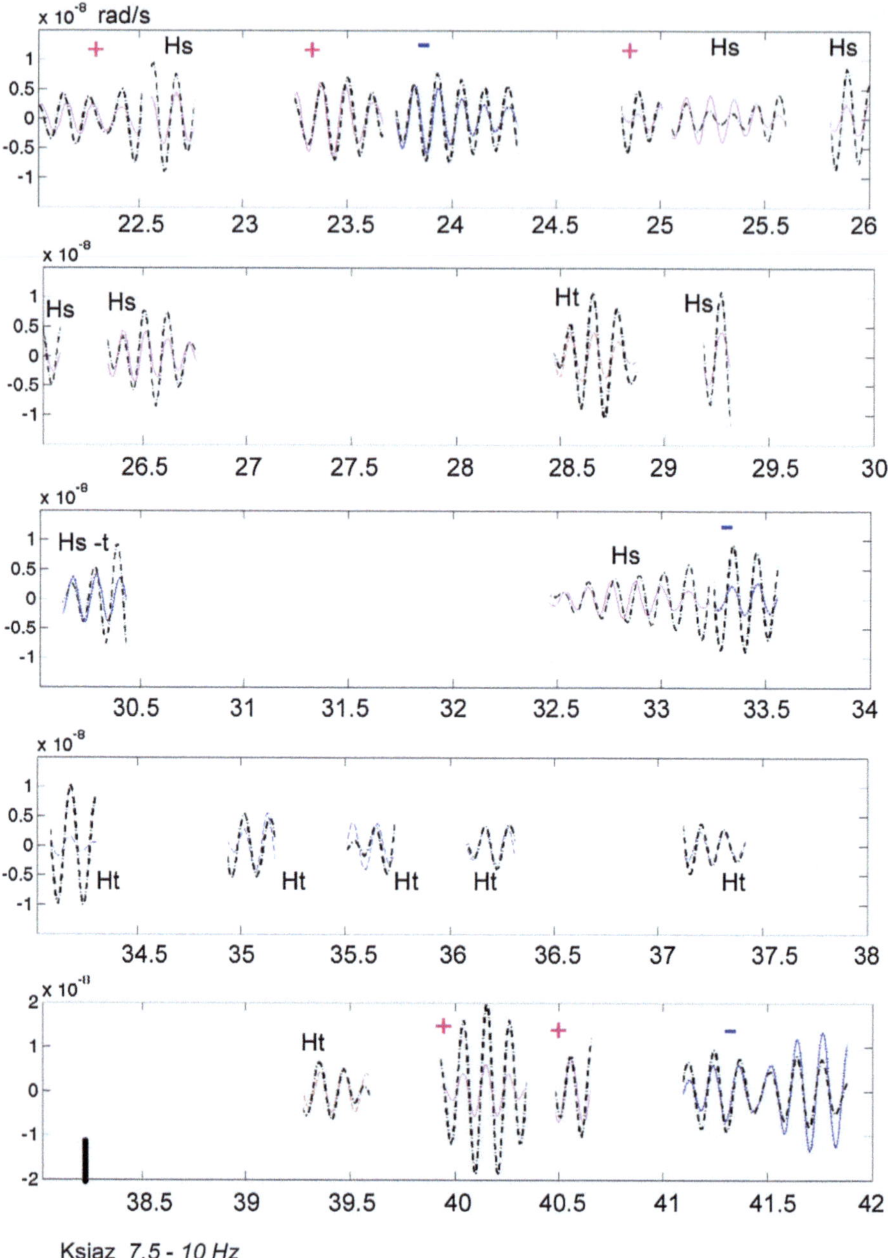

Fig. 4.3 Episodes of wave-groups conformity, 7.5–10 Hz. Analysis of recordings from rotational seismometers in Książ. Horizontal scale: time after start of the analysis time-period

Generally, records from Ojców and Książ in the 7.5–10 Hz band abound in conformant wave groups; in Ojców the main type of conformity is the "+" type: of twist with the spin; P waves onset also brings episodes of such a conformity. Greater variability is found in Książ, where "−" type conformance prevails, especially after S-wave arrival. Also in frequencies 3–7.5 Hz many congruent wave groups are found in the Ojców record, and again the "+" type conformance prevails. They are more varied in the recording from Książ in these frequencies, and less numerous before the start of shock.

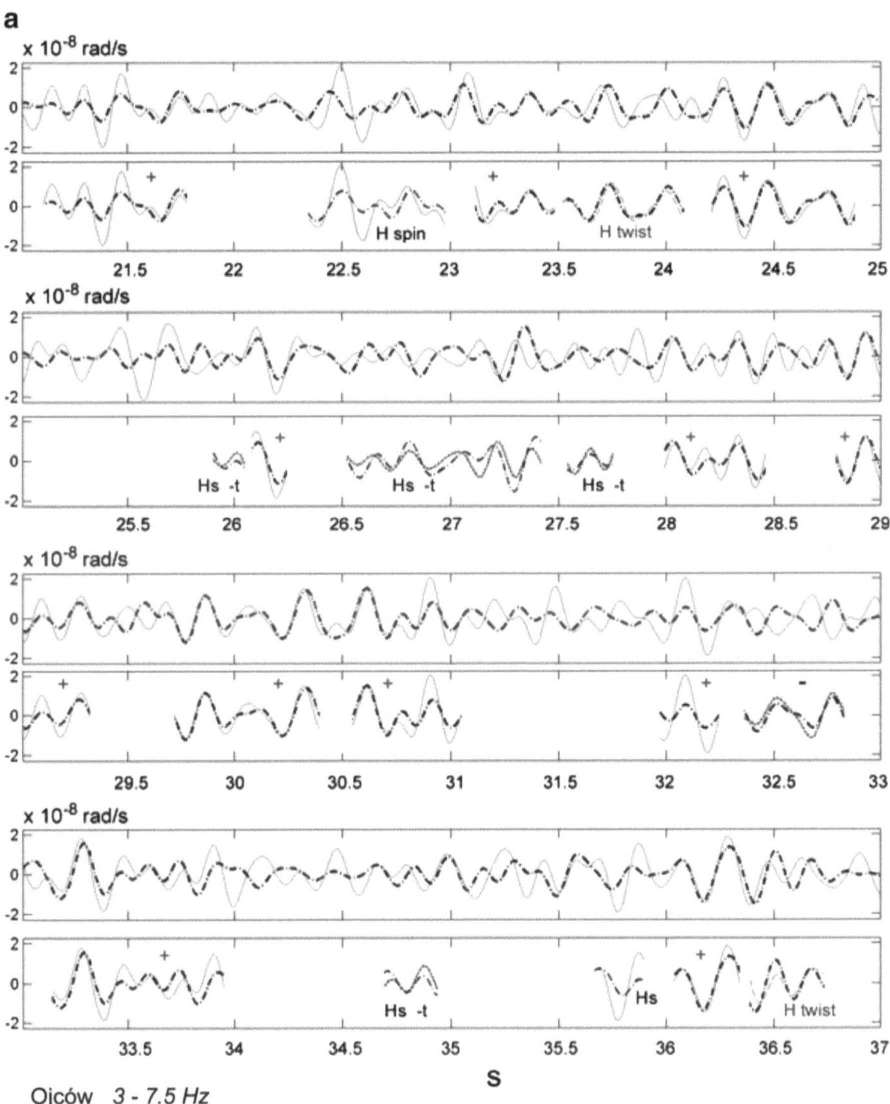

Fig. 4.4 continued

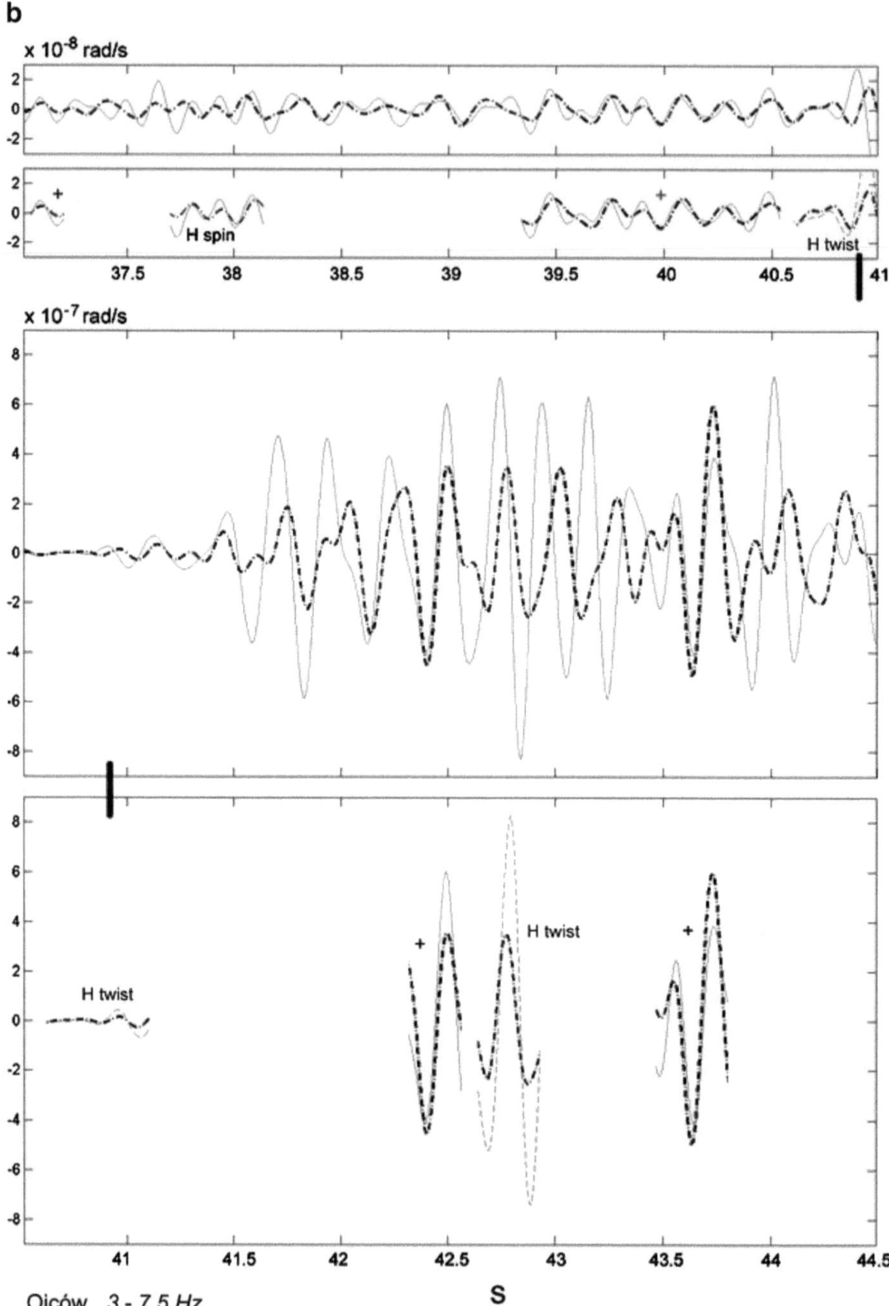

Fig. 4.4 Episodes of wave-groups conformity, 3–7.5 Hz. Analysis of recordings from rotational seismometers in Ojców: (**a**) first part; (**b**) second part including the onset of *P* waves. Horizontal scale: time after start of the analysis time-period

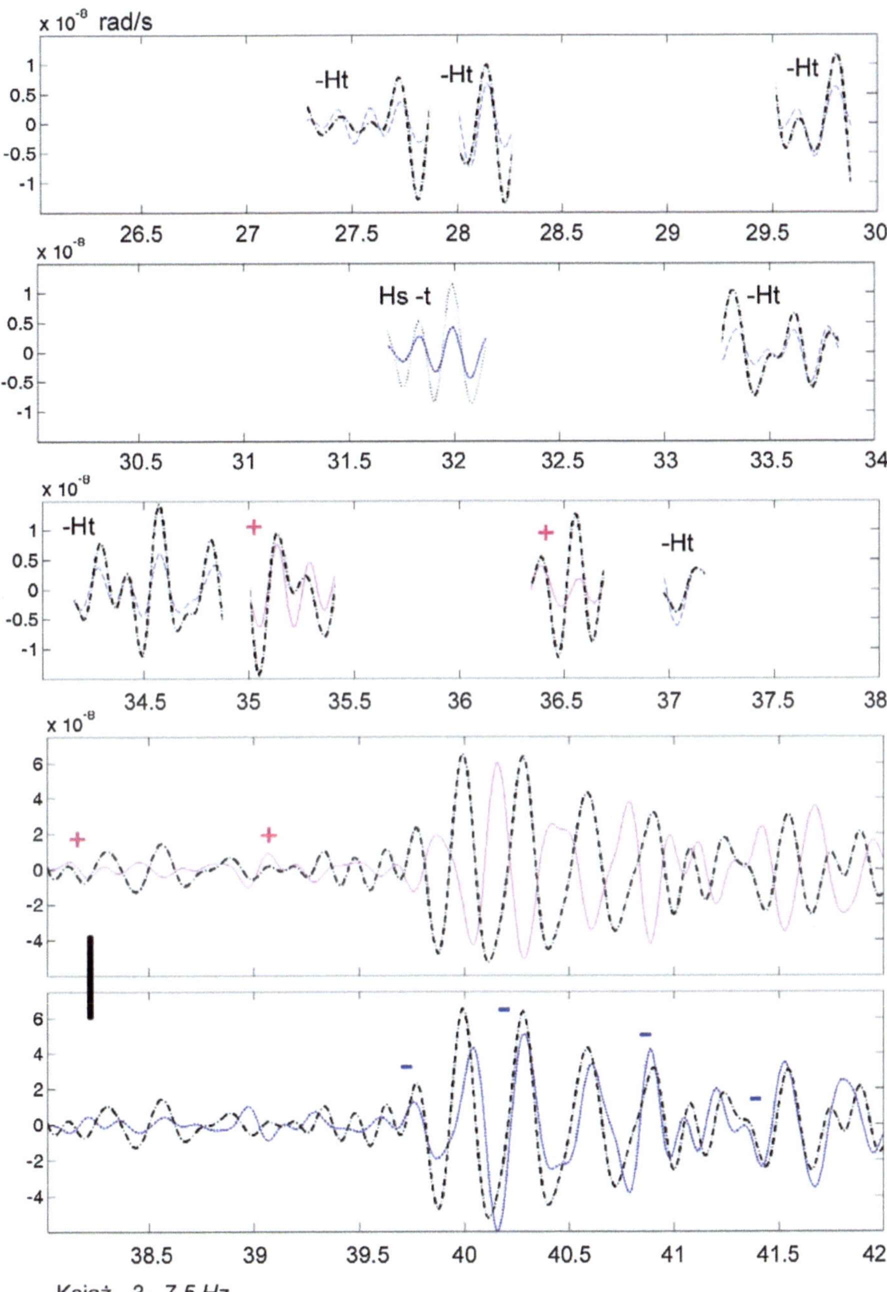

Fig. 4.5 Episodes of wave-groups conformity, 3–7.5 Hz. Analysis of records from rotational seismometers in Książ. Horizontal scale: time after start of the analysis time-period.The last time-period shown, which includes arrival of *P* waves, is represented twice; in the lower part of the figure – the first diagram shows curves of spin and twist, the second – the lowermost – spin and reversed twist. The *black vertical bar* marks the apparent moment of *P* waves arrival

The problem of inter-station signal concurrence is not banal here, as the event has been recorded in Książ later than in Ojców, because of the greater distance from the hypocenter. For the same reason, trace of the seismic event is much longer in the records from Książ. Therefore, some ways had to be found to properly align adequate diagrams. We used two kinds of markers for this alignment: the apparent P-waves arrivals found in the given frequency band and the prominent peaks visible on the diagrams of rotational components at this frequency band. We have settled with approximate alignment by prominent peaks, though it did not fit perfectly to the alignment by P-wave arrivals. Nevertheless, this non-perfect alignment by peaks is still correct in our opinion; some discrepancies should be attributed to differences in velocities and spectral characteristics of two seismic fields.

Both fields had emerged from one common source, but were recorded at two locations, lying at different directions and distances. Records at both stations contained numerous traces of rotational motions in the last 20 s before visible onset of the shock. There is possibility that these motions were generated by preparatory processes in the seismic focal zone, before the onset of shock.

We tried to find also the inter-station wave groups conformity in the rotational components, but results are not convincing yet. The same holds true for some other mining seismic events. Clearly, big distance between both observatories makes such studies difficult.

Pure P wave should not bear rotations; rotational waves should therefore travel as components of the slower, S waves. Therewith, it would be more reasonable to relate the occurrence of the first rotational waves conformances to the arrival of S waves. In most cases, however, P waves arrive first; nevertheless, they are accompanied by rotational components and the same was found in this study.

4.5   Discussion

An analysis of processed seismograms yielded visual evidence that the recorded rotational motions are related to processes in the focal zone. Conformance between rotational wave groups was already found in earthquake studies (Teisseyre 2011). The time-periods at the beginning of shocks and even preceding them appeared to contain the most interesting material, likewise in the present article.

We are aware that all these findings should be checked by an analysis of numerous cases and the use of adequate statistics – only then the conclusions could be accepted, rejected or modified.

The seismic noise is of course not purely seismic in its origin; many mechanical phenomena contribute to it. Therefore, the most desired findings during panning the pre-P-arrival part of data are those pairs of wave groups which include shifted component (Ht, −H twist and Hs). Such paired phenomena are likely to be produced in dynamic processes in rocks where the stresses are high; while direct similarity, e.g. of twist to spin, may be the effect of some other, casual incidents.

Distribution of signals into frequency bands seems justified and useful, as it helped to compare the waveforms. The effects of searching for interrelated rotational wave groups are somewhat unsatisfactory in the frequencies below 7.5 Hz, as it is hard or impossible to find really matching pairs. We attribute these difficulties to two causes: (1) all the frequency bands below 7.5 Hz were too wide; (2) in the frequencies below 2 Hz, the responses of the seismometers SM-3 vary more than in higher frequencies; therefore the rotational components are seriously distorted, and probably not in the same way on both stations.

The topic of conformant waves, and the rotation seismology in general, would benefit from simultaneous recording of the same events at two or more rotational seismic stations, lying at the same or at different direction from the source. This will be the way to discover relations between the distance and the detected data, a way on which classical seismology is far ahead. Also the influence of seismic ray path would therefore be investigated.

Studies on these relations should allow to cognize also a relationship between rotation and translational waves in various kinds of media and various types of the focal mechanisms. Such enquiries were already performed (Teisseyre et al. 2003a). The mining quakes reported there had produced strong rotational waves, but the share of rotation motions in radiated energy appeared low. Amplitudes of spin, twist and classic seismic signals (horizontal components) were also compared for a group of seismic events (Teisseyre 2011).

It should be mentioned that in the research on rotation/tilt components of the seismic field caused by frequent small ice-quakes on Alpine glacier (Górski and Teisseyre 2006), some unexplained undulations were observed shortly before and after shocks. These may indicate preparatory processes in ice, for example changes in creep rate. From comparison of initial and rotation/tilt signals of the ice-shocks recorded on Alpine glacier Pasterze and near the Hans Glacier on Spitsbergen, it appears that amplitudes in the latter are much greater than those in the small Alpine glacier. The difference was of several magnitude orders! (Teisseyre et al. 2004). Besides, rotation/tilt oscillations were unexpectedly strong on the Hans Glacier. Shear deformations and rotational motions may play a role in many preseismic and coseismic processes, valid for the earthquake study (see, Teisseyre 2008).

Rotation motion/waves study may lead to improved discerning between different kinds of rupture. Microscopically, shear fracture is a main fracturing process – even under confining pressure, the induced opposite shear centres help to fracture a material (Teisseyre et al. 2008). But the rotation motions play a role too – shear microdeformations and microfracturing are arranged spatially in the form of rotations. Therefore, considerable portion of released energy radiates as the rotational oscillations, probably of various orientations.

Under external shear conditions, organized shear deformation grows in the expanding flat zone which points out the forthcoming rupture. This zone wrestles with similar, transverse zones through the conflict of microspin motions. The share of rotational motions in the released energy should be low in this case, especially during the seismic event.

4.6   Conclusions

The results seem to support the Asymmetric Continuum Theory, according to which the rotational motions interplay with deformations and stresses in the preparatory stage of seismic events and during rupture, and, as a result, part of seismic energy is radiated in the form of rotational waves.

It is probable that certain small-amplitude motions, including rotational ones, occur in the focus during few seconds before the rupture starts, and these motions' echo is transmitted to the detecting station, where it is blended with other sources of the seismic noise. Nevertheless, it is sometimes possible to reveal these tiny motions by an analysis of several synchronous signals. These motions are candidates for immediate precursors.

References

Cochard, A., Igel, H., Schuberth, B., Suryanto, W., Velikoseltsev, A., Schreiber, U., Wassermann, J., Scherbaum, F., and Vollmer, D. (2006), *Rotational motions in seismology: theory, observation, simulation.* In: Teisseyre, R., M. Takeo and E. Majewski (eds.), *Earthquake Source Asymmetry, Structural Media and Rotation Effects*, Springer Berlin Heidelberg New York, 391-412.

Górski, M., and Teisseyre, K.P. (2006), *Glacier motion: seismic events and rotation/tilt Phenomena.* In: Teisseyre, R., M. Takeo and E. Majewski (eds.), *Earthquake Source Asymmetry, Structural Media and Rotation Effects*, Springer Berlin Heidelberg New York, 199-215.

Jaroszewicz, L.R. Krajewski, Z, Kowalski, H., Mazur, G., Zinówko, P., and Kowalski, J. (2010), *AFORS – Autonomous Fibre-Optic Rotational Seismograph: Design and Application*, Acta Geophysica vol. 59, no. 3 (2011), pp. 578–596

Moriya, T., and Teisseyre, R. (2006), *Design of Rotation Seismometer and Non-Linear Behaviour of Rotation*, in: Teisseyre, R., M. Takeo and E. Majewski (eds.), *Earthquake Source Asymmetry, Structural Media and Rotation Effects*, Springer Berlin Heidelberg New York, 439-450.

Teisseyre, K.P. (2008), *Charged dislocations and sources of electric field excitation.* In: Teisseyre, R., H. Nagahama and E. Majewski (eds.), *Physics of Asymmetric Continuum: Extreme and Fracture Processes*, Springer-Verlag Berlin Heidelberg, 137-159.

Teisseyre, K.P. (2011), *Spin and Twist Motions in the Earthquake Preparation Processes: Analysis of Records*, Acta Geophys., 59, 1, 5-28.

Teisseyre, K.P., Górski, M., and Suchcicki, J. (2004), *Seismic events and rotation waves in the Hans Glacier, Spitsbergen and the Pasterze Glacier, Austria*, Acta Geophys. Pol., 52: 4, 465-476.

Teisseyre, R. (2009), *Tutorial on New Developments in the Physics of Rotational Motions*, Bulletin of the Seismological Society of America, 99, No 2B, 1028-1039.

Teisseyre, R., and Górski, M. (2008), *Introduction to asymmetric continuum: fundamental point deformations.* In: Teisseyre, R., H. Nagahama and E. Majewski (eds.), *Physics of Asymmetric Continuum: Extreme and Fracture Processes*, Springer-Verlag Berlin Heidelberg, 3-16.

Teisseyre, R., and Górski, M. (2009), *Fundamental Deformations in Asymmetric Continuum*, Bulletin of the Seismological Society of America, 99, No 2B, 1132-1136.

Teisseyre, R., Górski, M., and Teisseyre K.P. (2008), *Fracture processes: spin and twist-shear coincidence.* In: Teisseyre, R., H. Nagahama and E. Majewski (eds.), *Physics of Asymmetric Continuum: Extreme and Fracture Processes*, Springer-Verlag Berlin Heidelberg, 111-122.

Teisseyre, R., Teisseyre, K.P., Moriya, T., and Palangio, P. (2003a), *Seismic rotation waves related to volcanic, mining and seismic events: near-field and micromorphic motions*, Acta Geophys. Pol., 51: 4, 409-431.

Teisseyre, R., Suchcicki,J., K.P. Teisseyre, K.P., and Palangio, P. (2003b), *Seismic rotation waves: basic elements of theory and recording*, Annals Geophys., 46: 4, 671-685.

Wiszniowski, J., and Teisseyre, K.P. (2010) *A Method to Minimize Errors in Seismic Differential Measurements* Acta Geophys. 58, 604-623

Chapter 5
An Analysis of Amplification Effects at Selected Polish Seismic Stations

Monika Dec

Abstract This paper presents an analysis of a seismic response of shallow geological structure beneath seismic stations in Książ (KSP), Ojców (OJC) and Górka Klasztorna (GKP) by means of the HVSR (Horizontal to Vertical Spectral Ratio) technique. This approach, developed by Nakamura (Q Rep Railway Tech Res Inst 30:25–33, 1989), allows to estimate the amplification effects due to coherent summing up of the direct and multi-reflected waves, while analysing spectra of horizontal and vertical parts of seismograms. This preliminary attempt to estimate the amplification effects is based on an analysis of seismograms from both mining induced and teleseismic earthquakes. The results show that amplification factor is less than 2 for all considered stations, suggesting very weak site effects. The HVSR curves for induced seismicity and for natural earthquakes are similar but not the same. The difference is visible at low frequencies, where the amplification effect estimated on the basis of local seismic records is apparently larger than for the teleseismic events.

5.1 Introduction

Site response analysis is part of assessing seismic hazard in earthquake-prone areas, since damages observed for destructive earthquakes are mostly associated with seismic wave amplification due to local site effects. Thus, there is a strong need for an even approximate estimation of seismic amplification effects for some critical locations. One of the techniques which allows to achieve this goal is the approach based on analysis of horizontal to vertical spectral ratios (HVSR) of ambient seismic records. This approach, often referred to as the Nakamura's technique (Nakamura 1989), has gained a large popularity (Frej 2008; Konno and Ohmachi

M. Dec (✉)
Institute of Geophysics, Polish Academy of Sciences, Księcia Janusza 64, 01-452 Warszawa, Poland
e-mail: monikadec@igf.edu.pl

A.F. Idziak and R. Dubiel (eds.), *Geophysics in Mining and Environmental Protection*, Geoplanet: Earth and Planetary Sciences 2, DOI 10.1007/978-3-642-19097-1_5,

1998; Olszewska and Lasocki 2004; Tsuboi et al. 2001), due to its simplicity and low cost of application. This method can be used to evaluate some characteristics of soft sedimentary deposits or applied to seismic microzonation investigations. The big advantage of HVSR technique is that it can be used in areas of low seismic activity, like Poland.

The aim of the presented analysis is to estimate an influence of the shallow subsurface structures beneath the broadband seismic stations belonging to Polish Seismological Network (PLSN) on recording seismograms. From nine digital seismic stations operating in Poland in the PLSN seismic network three stations (KSP, OJC and GKP) were selected for the amplification analysis.

5.2 Method

Figure 5.1 shows a typical geological structure of a sedimentary basin. Seismic waves passing through such a structure reflect many times beneath boundary layers, what results in a coherent summing them up and amplification of vibration at the surface. To evaluate this effect the spectrum of the vertical and horizontal components of vibrations is considered. Horizontal and vertical spectra on the surface and the bottom of sedimentary basin (H_f, V_f) can be written as:

$$H_f = A_h \cdot H_b + H_s, \quad V_f = A_v \cdot V_b + V_s,$$

where H_f and V_f are the horizontal and vertical spectra on the surface ground of the sedimentary basin; A_h and A_v are the amplification factors of horizontal and vertical motions of vertically incident seismic waves; H_b and V_b are the spectra of horizontal and vertical motion in the basement beneath the sedimentary layer; H_s and V_s are the horizontal and vertical spectra of Rayleigh waves.

In the sedimentary layer, the vertical component is assumed not to be amplified around the frequency range where horizontal component receives large amplification (Nakamura 1989), so $A_v = 1$. If the Rayleigh waves are disregarded,

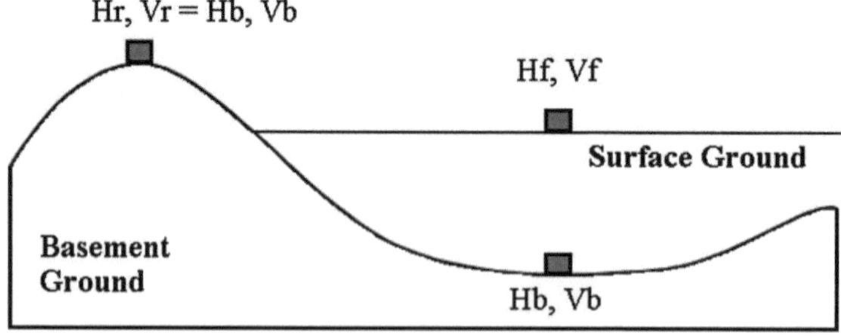

Fig. 5.1 Sketch of a typical geological structure of sedimentary basin (Nakamura 2000)

$V_s = H_s = 0$, and as a consequence, $V_f = V_b$. Usually, the vertical and horizontal components of the incident wave are comparable, $H_b \approx V_b$, so the ratio of horizontal to vertical movement components approximates the amplification factor:

$$HVSR = \frac{H_f}{V_f} = \frac{A_h\, H_b + H_s}{A_v\, V_b + V_s} = \frac{H_b}{V_b}\, \frac{\left[A_h + \frac{H_s}{H_b}\right]}{\left[A_v + \frac{V_s}{V_b}\right]} \approx A_h.$$

To calculate H/V spectral ratio as a function of frequency, the J-SESAME software was used (J-SESAME 2004). First, it calculates the horizontal and vertical spectra and then it calculates their ratio separately for two horizontal (NS, EW) components. In the next step, these two solutions are averaged. The maximum value of the HVSR is taken as an estimator of the amplification coefficient A_h, and the frequency at which it appears is interpreted as the resonance frequency of sedimentary layer. Nakamura ascertained that existence of the characteristic peaks at HVSR curves is strictly related to local geology, being independent of source parameters and signal frequency. The frequency range in which the HVSR method gives correct estimates of the amplification factor is still under investigation (Tsuboi et al. 2001; Olszewska and Lasocki 2004). This range depends on the structure of the subsurface layer but generally should not exceed 6–8 Hz (Olszewska and Lasocki 2004). As follows from Atkinson and Cassidy (2000), if the surface layer structure is relatively simple and HVSR exhibits a single peak, the value of the amplification factor obtained from the HVSR method correlates well with values provided by other surface methods. In more complicated geological situations, the HVSR method performs better (Atkinson and Cassidy 2000). Finally, the HVSR curves can be estimated for both P and S waves separately or together. In this study, both approaches were used when analysing seismograms from teleseismic events. In a case of local mining-induced sources, the amplification effects were calculated mainly from the S phases.

5.3   Data

The amplification effects were analysed for three seismic stations: KSP, OJC and GKP. Their location is shown in Fig. 5.2. Seismic observatory in Książ (KSP) exists since 1971 but digital broadband recording started there in 1997. This observatory is equipped with STS-2 seismometer located 42 m underground on hard rock, generally greywackes, so it is expected that site effect does not exist or is very small. For the second analysed seismic station, OJC, the seismometer is situated in a cave on about 200 m thick limestone layer. The digital recording commenced there in 1992 but broadband recording started in August 1999. The station is equipped with the STS-2 seismometer. The third analysed station, GKP, is also equipped with STS-2 seismometer. Located on thick soft sedimentary layers, it is expected to

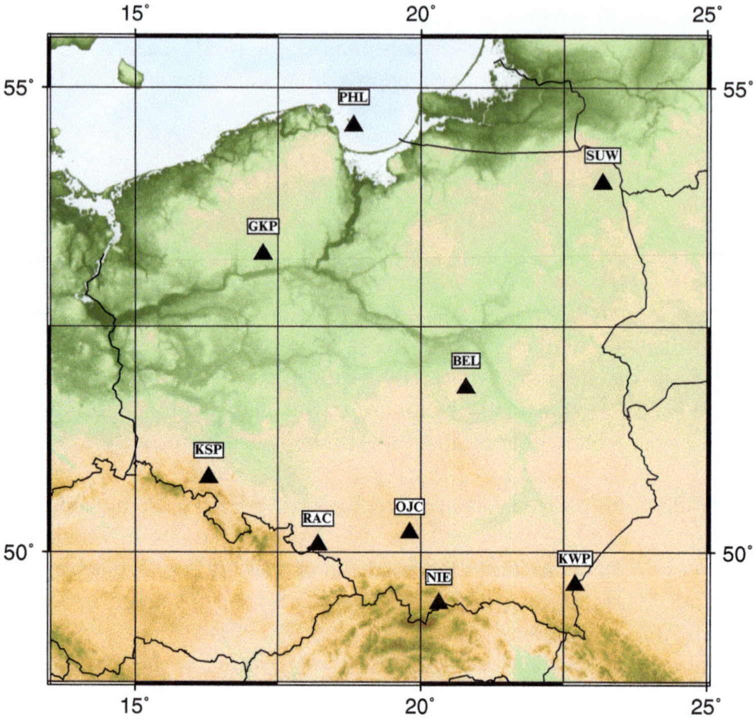

Fig. 5.2 Location of the Polish Broadband Seismic Stations run by the Institute of Geophysics, Polish Academy of Sciences

exhibit large amplification effects. All observatories are equipped with MK-6 seismic stations with recording dynamics of 140 dB and pass band from $93 \cdot 10^{-2}$Hz to 40 Hz. The amplification effects were estimated on the basis of seismic recordings of teleseismic and local events. While the teleseismic sources were essentially the same for all the stations, the local ones were different. In the case of KSP and GKP, the recordings of seismic events from the Lubin Copper Mine area were used. In the case of OJC, the local sources came from the Upper Silesian Coal Mine area. The seismograms used for the analysis were recorded with a sampling frequency of 100 Hz.

The number of events used was different for each stations. In the case of KSP station, 249 teleseismic and 276 local seismic events were analysed. Similarly, for OJC station 254 teleseismic and 242 local records were used. Finally, an analysis of the amplification effect for the GKP station was based on 125 teleseismic and 61 local events. For teleseismic sources, the earthquakes with magnitude greater than $M_b > 5.5$ were used. In the case of local sources, seismograms from rockbursts with magnitude above 2.0 in the case of KPS were considered. In the case of GKP, located further away from LGOM, the events with $M_L \geq 3.5$ were used. Finally, mining-induced tremors with $M_L \geq 2.2$ were analysed at OJC station.

5.4   Results

The Nakamura technique assumes accelerometric data. However, the Polish seismic stations are running velocity seismometers, so firstly we verified if the results of HVSR calculation based on velocity seismograms and their derivatives (accelerogram) are compatible. As it is shown in Figs. 5.3 and 5.4, there is no visible difference between both approaches, so in the next steps only original velocity seismograms were taken into account. Figures 5.4 and 5.5 show the HVSR curves calculated for KSP using local sources (Fig. 5.4) and teleseismic ones (Fig. 5.5). A very weak peak is visible in both cases but at different frequencies: 3 Hz for local events and around 4.5 Hz for teleseismic ones. However, in the former case the maximum peak value slightly above 1 suggests that the visible peak has no physical meaning and the amplification is roughly speaking constant in the considered case. It can be interpreted as a signature of non-existence of soft sediments below the KSP station.

The HVSR curves calculated for the OJC station are shown in Figs. 5.6 and 5.7. In both figures one can see a dominant peak at the same frequency of around 8 Hz. The peak maximum is about 1.7. In the case of mining sources (Fig. 5.6) one can see another well defined peak appearing at a frequency around 2 Hz. This peak and the peak maximum of 1.7 is also visible for the teleseismic events but its magnitude is slightly above 1. This comparison suggests that at small frequencies the local events are more amplified than the teleseismic ones.

Finally, the GKP station was considered. As it is seen in Figs. 5.8 and 5.9, the number of analysed events is much smaller than in the KSP case and OJC stations.

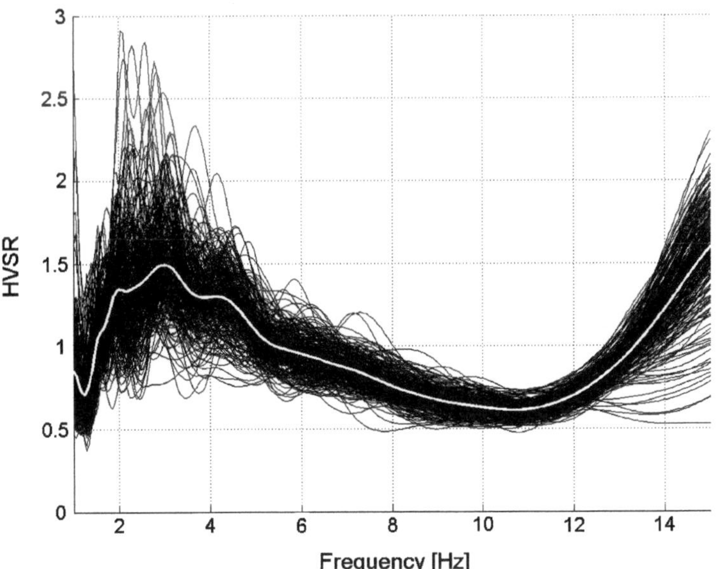

Fig. 5.3 The HVSR curves calculated for KSP station on the basis of original broadband velocity seismograms. *Black lines* are HVSR curves for particular mining-induced events. The *white line* denotes the average curve estimate for all 276 analysed events

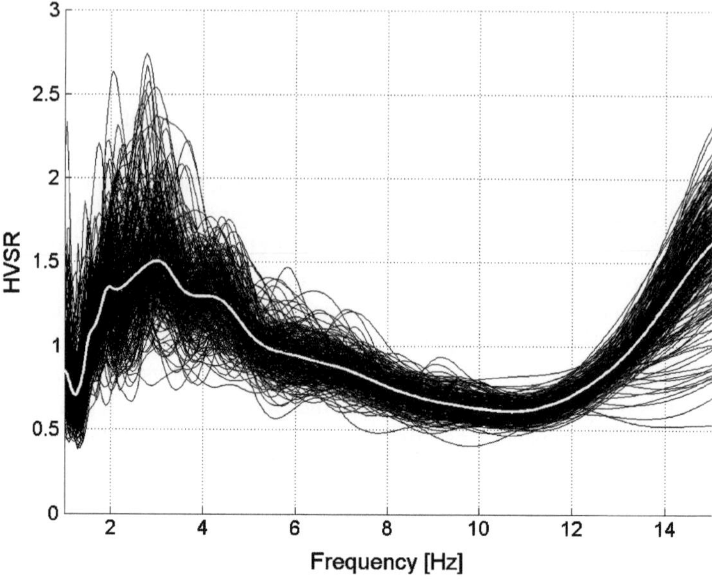

Fig. 5.4 The HVSR curve calculated for the KSP station using the derivative of original velocity seismograms. Comparing with Fig. 5.3 it is observed that both velocity and accelerometric waveforms lead to very similar results as theoretically expected

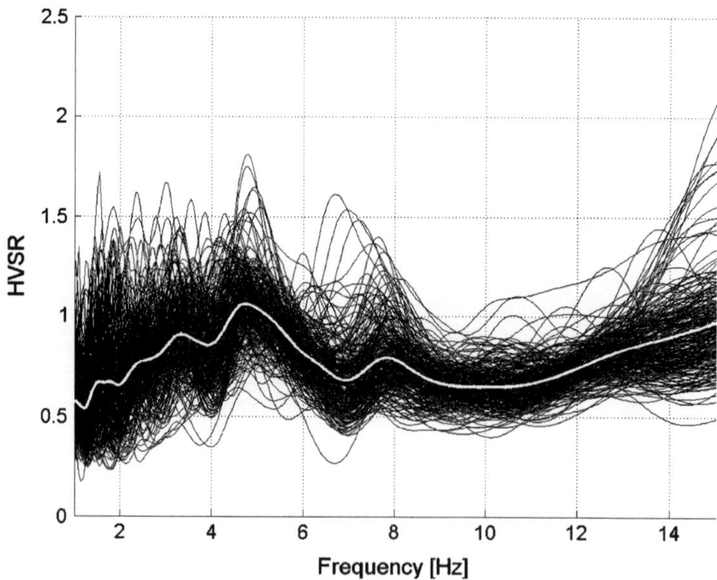

Fig. 5.5 The HVSR curves calculated for KSP station. *Black lines* are HVSR curves for particular teleseismic events. The *white line* denotes the average curve estimated for all 249 analysed events

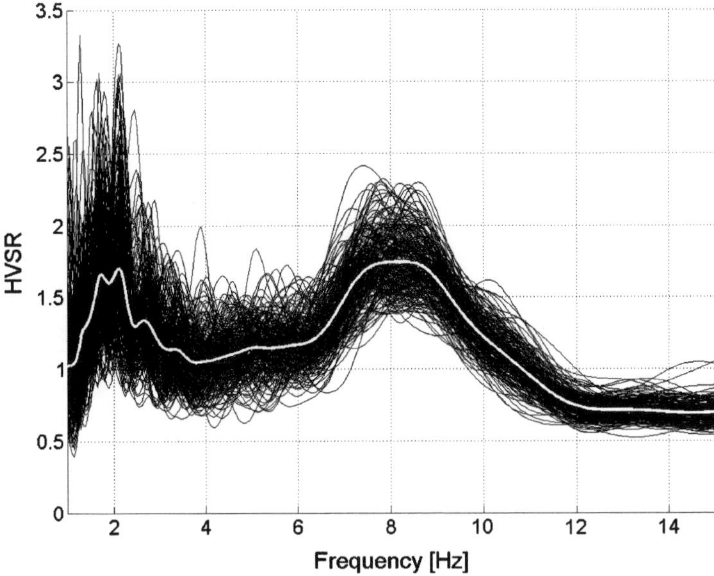

Fig. 5.6 The HVSR curves calculated for OJC station. *Black lines* are HVSR curves for particular mining-induced events. The *white line* denotes the average curve estimated for all 242 analysed events

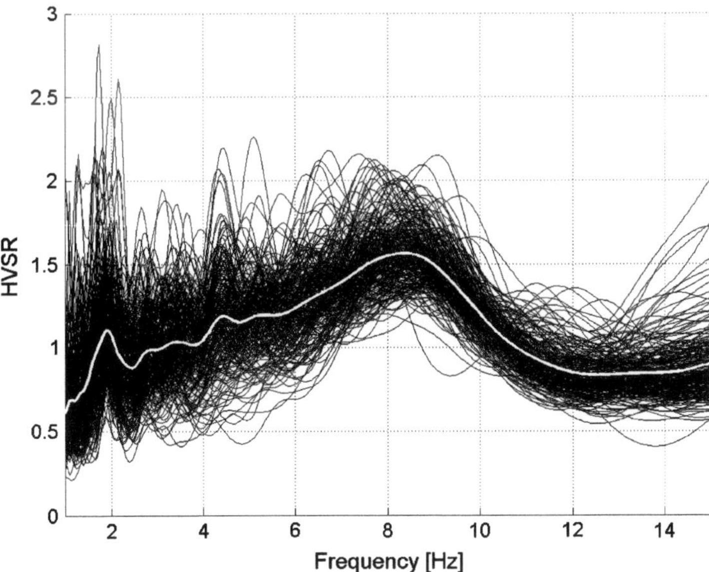

Fig. 5.7 The HVSR curves calculated for OJC station. *Black lines* are HVSR curves for particular teleseismic events. The *white line* denotes the average curve estimated for all 254 analysed events

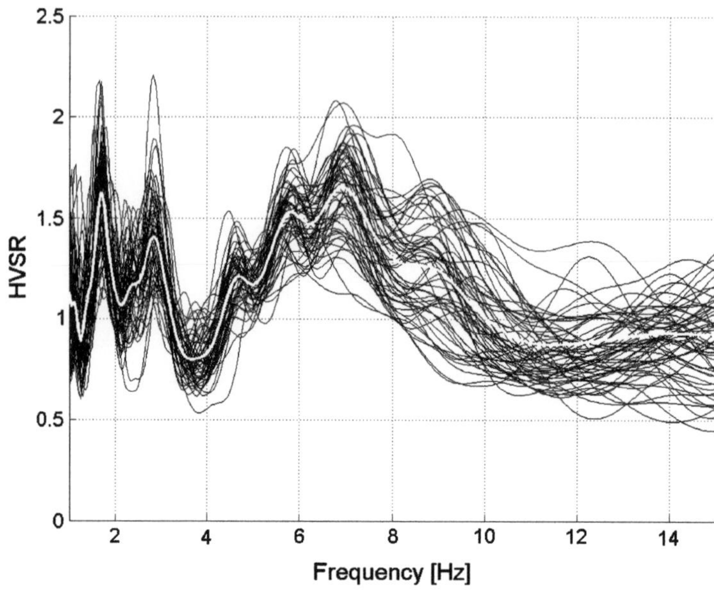

Fig. 5.8 The HVSR curves calculated for GKP station. *Black lines* are HVSR curves for particular mining-induced events. The *white line* denotes the average curve estimated for all 61 analysed events

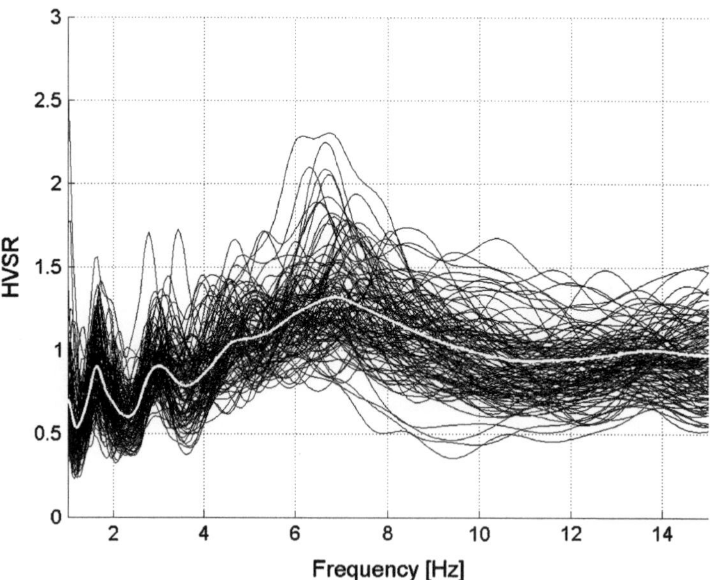

Fig. 5.9 The HVSR curves calculated for GKP station. *Black lines* are HVSR curves for particular teleseismic events. The *white line* denotes the average curve estimated for all 125 analysed events

However, for almost all the events considered, the HVSR curves coincide well, so one can expect reliable average results in this case. Similar to OJC and KSP, also for GKP one can notice the existence of the main peak of the HVSR curve, but now it occurs at a slightly lower frequency: around 7 Hz. Also its maximum is slightly lower and the average amplification reaches 1.5. A comparison of amplification for teleseismic and local events exhibits a significant difference at a frequency above 4 Hz. The local source based on HVSR curves exhibit two clearly visible peaks occurring at frequencies of 1.8 and around 3 Hz, and reaches the values of 1.6 and 1.4, respectively.

5.5 Conclusions

This work presents preliminary results of an analysis of amplification factors for three Polish stations in Książ, Ojców and Górka Klasztorna. For all these stations peaks in the H/V spectral ratio with the maximum values around 1.5 were observed. These maxima appear at frequencies from 2 to 8 Hz. The relatively flat shape of HVSR curves with H/V values oscillating around 1 in the analysed frequency range suggest that there are no significant amplification effects at KSP, OJC and GKP stations.

A comparison of the HVSR curves for local and teleseismic sources is very interesting. In the case of KSP and OJC stations the HVSR curves for local events exhibit lower values at frequencies below 4 Hz than for teleseismic sources. Indeed, at the local-based HVSR curves one can notice, in addition, an enhancement of H/V spectral ratios as a well defined peak, much less pronounced at teleseismic-based solutions. In the case of GKP station, the situation is slightly different. In this case the all HVSR curves are shifted towards higher values and all peaks are more pronounced. The results suggest that some amplification is visible at GKP site, while at OJC and KSP sites there is actually no amplification effects.

References

Atkinson, G. M. and Cassidy J. F. (2000). Integrated Use of Seismograph and Strong-Motion Data to Determine Soil Amplification: Response of the Faser River Delta to the Duval and Georgia Strait Earthquakes. *Bull. Seismol. Soc. Am. 90*, 1028–1040.

Frej, A. (2008). Amplifikacja drgań wywołanych wstrząsami górniczymi rejonie niecki bytomskiej. [Amplification of mining-induced vibrations in the Bytom region], Ph.D. Thesis Rozprawa doktorska – Uniwersytet Śląski (in Polish).

J-SESAME. (2004). User Manual Version 1.08. July 2004.

Konno, K. and Ohmachi T. (1998). Ground-Motion Characteristics Estimated from Spectral Ratio between Horizontal and Vertical Components of Microtremor. *Bull. Seismol. Soc. Am. 88*, 228–241.

Nakamura, Y. (1989). A Method for Dynamic Characteristics Estimation of Subsurface using Microtremor on the Ground Surface. *Quarterly Report of Railway Technical Research Institute, Japan. 30*, 25–33.

Nakamura, Y. (2000). Clear Identification of Fundamental Idea of Nakamura's Technique and Its Applications. In *proceedings of the 12th World Conference on Earthquake Engineering, Auckland, New Zealand, Volume 2656*, pp.1–8.

Olszewska, D. and Lasocki S. (2004). Application of the horizontal to vertical spectral ratio technique for estimating the site characteristic of ground motion caused by mining induced seismic events. *Acta Geophys. Pol. 52*, 301–318.

Seekins, L. C., Wennerberg, L., Margheriti, L., and Liu H.-P (1996). Site Amplification at Five Locations in San Francisco, California: A Comparison of S Waves, Codas and Microtremors. *Bull. Seismol. Soc. Am. 86*, 627–635.

Tsuboi, S., Saito M., and Ishihara Y. (2001). Verification of Horizontal-to-Vertical Spectral-Ratio Technique for Estimation of Site Response Using Borehole Seismographs. *Bull. Seismol. Soc. Am. 91*, 499–510.

Chapter 6
Investigation of the Medieval Jeroným Mine Stability: Present Results from a Distributed Measurement Network

Jaromír Knejzlík, Zdeněk Kaláb, Markéta Lednická, and Lubomír Staš

Abstract Jeroným is a sixteenth century mine situated in western Bohemia near Mariánské Lázně. It is being prepared to be the base of a museum of medieval ore mining. As a source of information for investigation of stability of aged mining openings, a distributed measurement network (DMN) combined with seismic station has been built up gradually since 2004. DMN is equipped with sensors for measurement of mine water level changes (3 selected monitoring points), mine temperatures (2 points), rock massif joint aperture changes (4 points), cross-section convergence changes (2 points) and changes of rock massif stress tensor (2 points). Data acquisition unit of DMN is integrated into the seismic recording station that is equipped with data transmission via GSM network. Installation of all above mentioned sensors was finished in May 2009. A brief description of DMN and present results derived from the data obtained are described in the contribution. Monitoring of height changes of high inaccessible ceiling using laser distance meter and changes of stress tensor of rock massif are described in detail.

6.1 Introduction

In several papers, e.g. Kaláb et al. (2006, 2008a) or Žůrek and Kořínek (2004), information on the location of Jeroným mine, brief history of mining activities in this place, the purpose and the aim of the present research of stability of mining

J. Knejzlík (✉) • L. Staš
Institute of Geonics of the Academy of Sciences of the Czech Republic, v.v.i., Studentská 1768, 708 00 Ostrava-Poruba, Czech Republic
e-mail: knejzlik@ugn.cas.cz; stas@ugn.cas.cz

Z. Kaláb • M. Lednická
Institute of Geonics of the Academy of Sciences of the Czech Republic, v.v.i., Studentská 1768, 708 00 Ostrava-Poruba, Czech Republic
and
VŠB-Technical University of Ostrava, 17. Listopadu 15, 708 33 Ostrava-Poruba, Czech Republic
e-mail: kalab@ugn.cas.cz; lednicka@ugn.cas.cz

A.F. Idziak and R. Dubiel (eds.), *Geophysics in Mining and Environmental Protection*, Geoplanet: Earth and Planetary Sciences 2, DOI 10.1007/978-3-642-19097-1_6,
© Springer-Verlag Berlin Heidelberg 2011

openings were described. Distributed measurement network (DMN), which is installed in the part of the mine called Abandoned Mine Workings (AMW), was also described therein. The current configuration of DMN, used sensors and additional hardware were described in the paper by Knejzlík and Rambouský (2008). As the underground space is visited only occasionally, DMN has to work as an independent system without attendance.

Using mechanical instruments, geotechnical monitoring is performed quarterly (Kaláb et al. 2008a). Since 2006, selected monitoring points have been monitored using DMN (1/h). Measurements results of galleries and chambers convergences and observing of cracks, especially on rock pillars, did not show any significant changes of recorded parameters. Therefore it is likely that rock massif has been stable at the monitored points during the time of monitoring. Strain gauges based on inductive displacement sensors are used in recent configuration of DMN, because the inductive sensors are more accurate and more stable (first version of DMN exploited displacement sensors with strain-gauge bridges). Results of seismic loading of Jeroným mine mainly during restoration of the drainage adit and during West Bohemia swarm in 2008 are described by Kaláb and Lednická (2006). Significant results are obtained also from monitoring of mine water level changes, for example, a sudden decrease of water levels on points near K2 chamber (Kaláb et al. 2008b or Kaláb and Lednická 2009).

In addition to the above-mentioned customary measurements, two new experimental measuring methods were incorporated to DMN:

- Measurement of the distance between the bottom and inaccessible ceiling of the large mining chamber (convergence of ceiling) using laser distance meter (LDM).
- Monitoring of changes of the stress state in rock massif using conical strain gauge probe (CCBM).

Both above-mentioned methods are tested for future monitoring of stability of K2 chamber during driving a connecting gallery between parts called Old Mine Workings and AMW, which are separated now. Places for operational installation of LDM sensor and CCBM probes (and also other sensors) are partly selected on the basis of mathematical modeling of stress field in the surrounding of K2 chamber (Hrubešová et al. 2007).

6.2 Measurement of Height of an Inaccessible Ceiling of Mining Chamber Using LDM

Chamber named K2 is a large space. Its ceiling is at a height of about 7 m above the bottom. The bottom and the ceiling have very rough surface and bottom can be flooded. Installation of an anchor of mechanical extensometer requires high scaffolding. However, it is possible to measure convergences of the ceiling height with permanent installed laser distance meter (LDM). In such a case, it is not necessary to install any equipment on the ceiling, as LDM is able to measure reflection from a rough surface. On the other hand, laser measurement has lower accuracy and resolution ability in comparison with mechanical measuring instruments.

Commercially available LDM sensors for industrial automation are known. As an example, we demonstrate the parameters of AccuRange 4000 Laser Rangefinder developed by Acuity Company (www.acquitylaser.com):

- Range of measurement: up to 16.5 m
- Accuracy: 2.55 mm (reflection from diffused surface)
- Resolution ability: 0.32 mm
- Cover type IP67
- Output: digital RS-232, optional analogue 4–20 mA.

This type of LDM would be advantageous for application in DMN. However, other commercially available equipments have similar parameters. Its high price is the main disadvantage.

In the Jeroným mine, laser measurements of chamber ceiling heights are performed by LDM type Leica DISTO – regular quarterly manual measurements. Based on this good experience we have decided to integrate Leica DISTO™ A4 to the DMN. The manufacturer states the following parameters:

- Range of measurement: 0.05–200 m (up to 100 m without reflecting target)
- Accuracy: ±1.5 mm
- Resolution ability: ±1 mm (0.1 mm in remote control mode).

LDM Leica DISTO™ A4 uses simple text protocol for control instructions and measured data transfer. For the purpose of its use as an intelligent measuring unit in DMN, it was necessary to evolve electronic interface. This interface contains a microprocessor with a compiler of communication protocols Aibus2 – Leica DISTO™ A4 and circuit for feeding of Leica DISTO™ A4 from DMN bus.

LDM Leica DISTO™ A4, together with the interface, is installed in a waterproof case which is equipped with visor and body tube on the foreside. The case is mounted on the modified parabolic satellite antenna tripod – see Fig. 6.1. The body tube is used as a protection against drip water from the ceiling; vertical measurement is not possible in the mine due to dropping water. The foot-wall of K2 chamber consists of blocks of rock, deposits of mud and flowing or stagnant mine water (Fig. 6.2).

Tripod of LDM was mounted on the roughly horizontal outcrop of granite. Basement for horizontal supports of tripod legs was leveled to the horizontal position. Screw bolts for fixation of horizontal supports were embedded to concrete in holes bored to the rock. All the horizontal supports were buried to concrete after bolting. The rod of the tripod has been adjusted to the vertical position after hardening of concrete. The laser was directed at approximately uniplanar plane on the ceiling, which is perpendicular to its beam direction (12° from vertical). Diameter of the laser beam at reflecting surface is about 8 mm. Measurement was started on 14 May 2008. Sampling interval was set to 1 h. Chart of instantaneous

Fig. 6.1 Interface with LDM Leica DISTO™ A4 mounted on the modified support of the parabolic satellite antenna

Fig. 6.2 LDM Leica DISTO™ A4 installed in K2 chamber

values and moving average (24 h) of data recorded by LDM since 14 May 2008 until 12 July 2009 are presented in Fig. 6.3.

Two significant gaps of data registration longer than 1 day occurred (since 31 July 2008, 12:00, until 4 August 2008, 13:00, and since 1 October 2008, 8:00, until 6 October 2008, 5:00). The first one was caused by a software error, the second one by overfilling the disk capacity due to unexpected extreme amount of recorded seismic events during the occurrence of West Bohemia seismic swarm. There are some interesting observations gained from LDM data. Generally, medium measured value is $h = 5{,}912$ m and whole the data lies within the interval of declared accuracy of LDM Leica DISTO™ A4 (± 1.5 mm) with the exception of

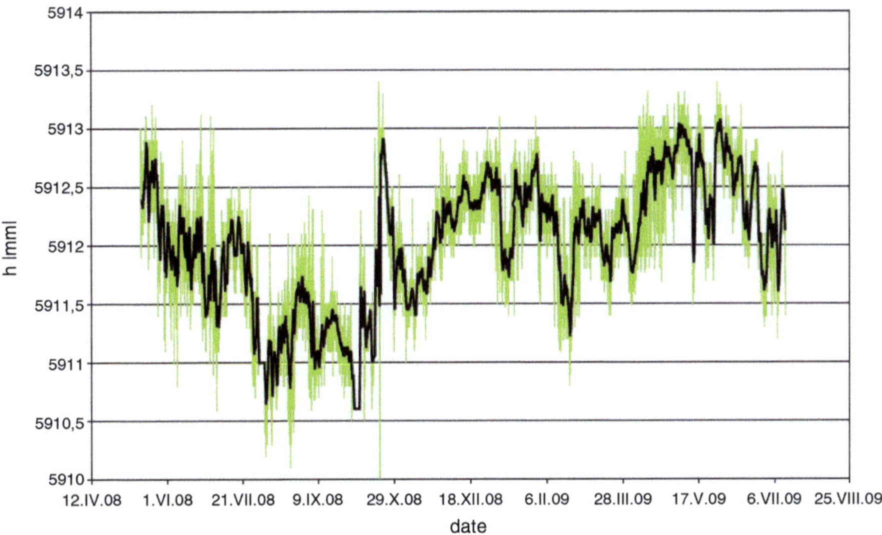

Fig. 6.3 Detail of slant distance *h* between LDM and the ceiling. Date interval from May 2008 to July 2009. Extreme values from 19 October 2008 are removed. Instantaneous values – *green line*, moving average 24 h – *black line*

three values recorded on 19 October 2008. We suppose that those extreme values could be impressed by vibrations of tripod caused by seismic or human activity. It is possible to observe short-term and long-term variations of moving average of data and variations of dispersion of instantaneous values on the graph. They are probably caused by combination of real variations of distance evoked naturally, by measurement errors of Leica DISTO™ A4 and by temperature instability of the primitive tripod. Rough reflecting surface of the ceiling and its roughly perpendicular direction lowers the accuracy of measurement notably. Nevertheless, it is possible to find interesting changes in data trends. For example, a notable peak occurred during the seismic swarm in West Bohemia after 19 October 2008. Some trends of recorded data seem to be influenced by tidal cycles of the Earth, yet our data set is too short for verification of this hypothesis and measured values are interfered by different factors.

In the case of LDM, the main task is to verify a possibility of long term sufficiently accurate measurement without a reflecting target. It is possible to declare that presented design of LDM is sufficiently accurate and stable for planned continuous monitoring of stability of chamber K2 during driving a connecting gallery even without using a reflecting target.

More modern laser distance meter Leica DISTO™ A6 is now available in market. It is equipped with communication interface Bluetooth. We prepared interface for implementation to the DMN for this new type. However, its disadvantage is less resolution – 1 mm only.

6.3 Experimental Application of CCBM Probes in DMN

Conical strain gauge probes for measurement of stress in rock massif are based on the principle of evaluation of a sufficient amount of independent deformations of the conical bottom of the borehole using overcoring method (Nakamura et al. 1999; Kang 2000). We apply the CCBO modification of this method (Compact Conical-ended Borehole Overcoring). Methodology and design of the first generation of CCBO probes is described in the paper presented by Staš et al. (2005). The second generation of CCBO probes uses fail-safe communication protocol for communication between embedded microprocessor (internal data logger) and external computer – see Knejzlík et al. (2008). Using simple modification of communication interface and feeding of electronic circuits we developed a new version of a probe named CCBM (Compact Conical-ended Borehole Monitoring) from CCBO probe with the possibility of functioning as an intelligent measuring unit in DMN. CCBM allows to measure deformations from which it is possible to evaluate stress change. We use this type of probes for measurement of stress changes induced by long-wall progress in mines. In this case, data is measured manually with using of portable terminal (Staš et al. 2007).

CCBM probe was used for long-term monitoring with telemetric data transmission in Jeroným mine for the first time. It was necessary to solve following problems:

- Drilling of boreholes in medieval mine without any technological maintenance.
- Sticking of CCBM probe in conditions of low temperature and high moisture.
- Developing of CCBM probe interface which guarantees high separation of the probe circuits from DMN bus. This separation is important due to penetration of electric disturbances from DMN bus to highly sensitive circuits of the probe. Lightning protection is also very important because sensitive input circuits of strain gauge bridge can be damaged by danger voltage evoked by a stroke of lightning in the surroundings.

Trial installation of CCBM was done in K1 chamber. This location was separated to provide information about stability situation in one of the largest chambers in the mine, where waste rock material used to be stored. The possibility of installation of drilling machine and assuring power was the second reason. However, significant results will be obtained after starting measurement of stress changes in K2 chamber during driving the connection gallery. Two boreholes with diameter 76 mm have been drilled recently, slightly upraise (1–3°) to depth of approximately 3 m in walls of K1 chamber (Fig. 6.4). The CCBM1 probe was installed close to a shaft in the borehole oriented to 111° in the system NESW. The borehole on the CCBM1 position is situated in homogenous granite without cracks (according to examination of borehole using portable camera). CCBM2 was installed close to the portal of gallery connecting K1 and K2 chambers. The borehole is oriented to 262° in the system NESW. This borehole was drilled through a faulted zone and was ended after intact rock had been reached.

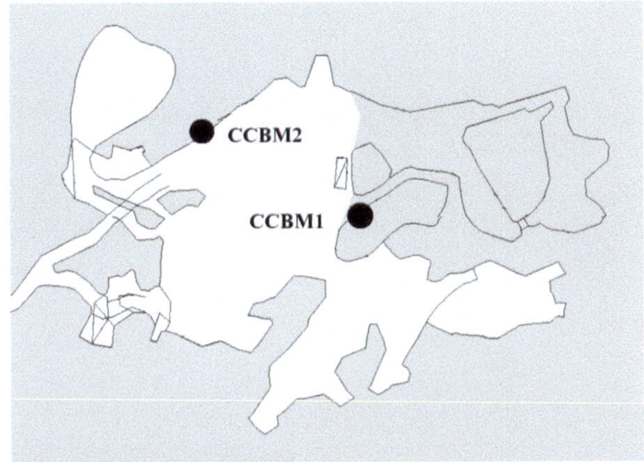

Fig. 6.4 Sketch of localization of CCBM probes in K1 chamber

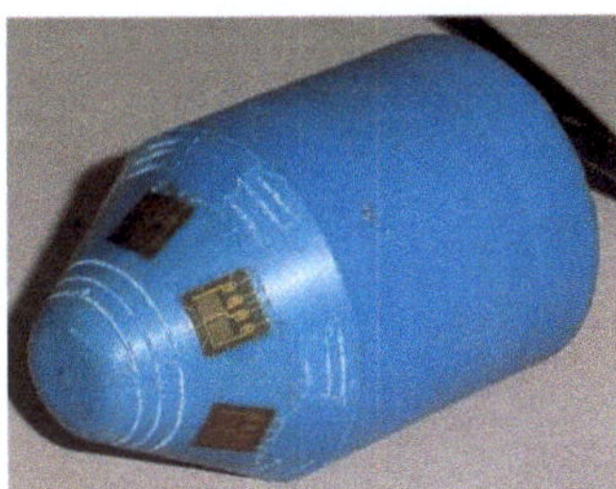

Fig. 6.5 Photo of CCBM probe

Fig. 6.6 CCBM1 probe installed in borehole

CCBM probe interfaces contain converter of communication standards RS485/ RS232 with galvanic separation between ports and DC/DC converter. Electronic circuits of CCBM probes are not connected to the ground connection.

Measuring tip of CCBM probe contains six strain gauge 2-element 90° tee-rosettes – see photo in Fig. 6.5. Longitudinal components are labeled SG1L-SG6L, transversal ones are labeled SG1T-SG6T.

Both CCBM probes were bonded to the boreholes in October 2007 (Fig. 6.6). A special two-part glue was applied for this. Exact orientations of probes, i.e. positions of SG1L strain gauge, were fixed using special instrumentation. Regular registration of both probes was started on 17 November 2007. The obtained data

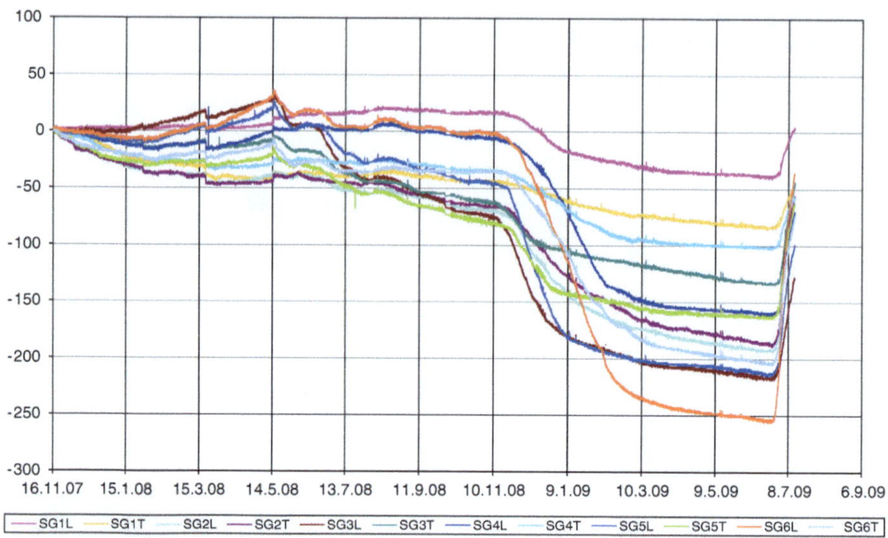

Fig. 6.7 Plot of measured strain changes from CCBM1 probe (scale factor on Y axis: μStrain)

manifest proper functioning of electronic circuit of both probes and their interfaces, low level of noise and good immunity against electric disturbances.

Data from CCBM1 probe are quite correct. It allows making initial interpretation of stress changes. Plot of all measured strain changes from CCBM1 probe is presented in Fig. 6.7. Little discontinuity in data row caused by irregular modification of software in DMN master unit occurred between 22 March 2008 and 15 May 2008. Only small changes of strain were registered before June 2008. This can be caused by glue hardening, perhaps. Interesting change of strain conditions from June 2008 can be observed. Compression grew on all strain gauges and it finished suddenly in July 2009. Initial computation was made for time interval from November 2008 to July 2009, when changes were significant. Theory of calculation is described in Kang (2000).

Mechanic properties of rock material (granite), i.e., Young's modulus of elasticity E and Poisson's ratio μ, were evaluated on the basement of laboratory tests of cores obtained from the above-mentioned boreholes. Dispersion of measured input parameters E and μ is large (8.6G Pa $\leq$ E $\leq$ 16.5 GPa; 0.11 $\leq$ μ $\leq$ 0.26). The values E $=$ 13 GPa and μ $=$ 0.16 were used for trial calculation.

Plot of calculated principal stress components (S1, S2 and S3) is presented in Fig. 6.8; gradual change of orientation of principal stress components is illustrated in Fig. 6.9. Maximum reached values are very low, less than 3.5 MPa. However, continuance of stress is very interesting, especially the short return towards tensile stress in July 2009. The result is more predicative for trends of the obtained value relations and trends of stress orientations than for absolute values of stress changes.

Interpretation of the natural reason of the stress changes measured is difficult due to a short data row and a very complicated structure of the rock massif. Current

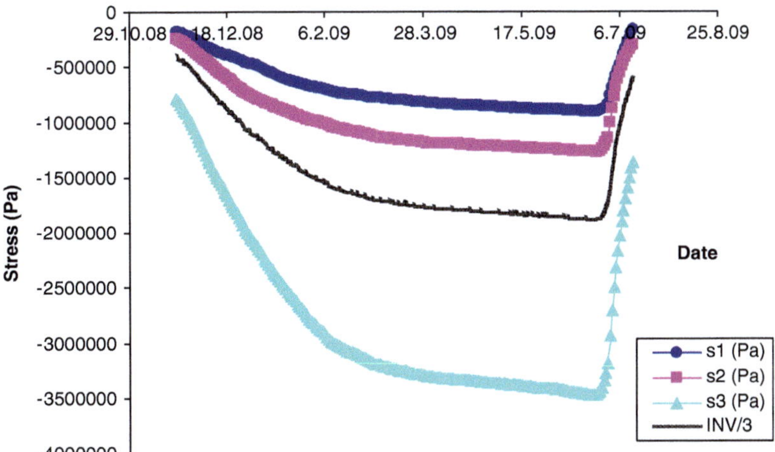

Fig. 6.8 Plot of principal stress components

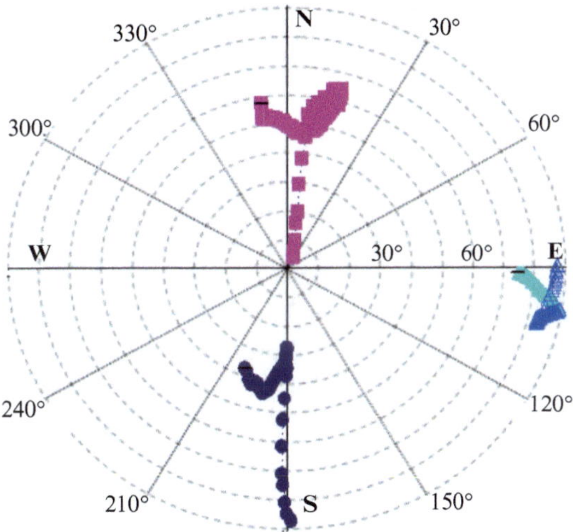

Fig. 6.9 Projection of principal stress components. (S1 – *circlet*, S2 – *square*, S3 – *triangle*; upper hemisphere – *void markers*, lower hemisphere – *filed markers*)

result indicates that the stress field in the surrounding of K1 chamber is not absolutely stable. Projection of principal stress components at the bottom and/or upper hemisphere is used in Fig. 6.9. The maximum compression stress component S3 is significantly horizontally oriented into east–west direction. However, there are very similar values of S1 and S2 in final sequences (Fig. 6.8), whence it follows that orientations of components S1 and S2 (Fig. 6.9) are not sufficiently representative.

Data records from CCBM2 probe have a different character. Generally, monotonous courses of tensile and/or compression on individual strain gauges are recorded. In October 2008 (during seismic swarm in West Bohemia), adjacent transversal strain gauges SG1T and SG6T broke down. Possible reason can be that the borehole for CCBM2 probe was drilled in more faulted rock. A micro-crack could have been omitted or occurred due to vibrations in the place where CCBM2 was installed.

6.4 Conclusions

To obtain more detailed information on the rock massif behavior, distributed measurement network for evaluation of hydrologic, geomechanical and other parameters has been built up in the Jeroným mine since the first half of 2006. DMN is integrated to the existing seismic recording station equipped with data transmission via GSM network to the registration centre in Ostrava. The complete monitoring system is constructed as modular so that it is possible to change the configuration of this system.

The instrumentation is generally based on a commercial control kit and sensors. New types of sensors for monitoring of mine cavities stability (CCBM probes, LDM) are tested. Experimental operation of LDM and CCBM probes proved possibility to be reliably operated within the scope of DMN and provided new technological findings for its design and installation. There are anomalies in data rows recorded by LDM and CCBM1 probe, which point at natural origin of the low rock massif stress state changes in the surroundings of Jeroným mine. Its connection with occurrence of the seismic swarm in West Bohemia can be neither proved, nor excluded. It is going to be the subject of research in the future.

Current measurements are performed in stationary conditions of a medieval mine (release condition). Data from quarterly observing and DMN document that monitored places have been stable until now. New findings can be expected after the reconstruction of the underground space starts in terms of preparation of a museum. Driving of the cross-connection gallery between two separated parts of Jeroným mine will be the most significant activity.

Acknowledgement This paper has been achieved with financial support of the Czech Science Foundation, project No. 105/09/0089.

References

Hrubešová, E., Kaláb, Z., Kořínek, R. and Žůrek, P. (2007) Geotechnical Monitoring and Mathematical Modelling in Medieval Mine Jeroným (Czech Republic). Górnictwo i Geoinżynieria, Vol. 31, Zeszyt 3, 183–190.

Kaláb, Z., and Lednická, M. (2006) Interpretation of Seismic Records of Blasting Operations Performed in Gallery Jeronym in Cista. Transactions, Civil Engineering Series, Vol. VI, No.2/2006, 155–160 (in Czech).

Kaláb, Z., Knejzlík, J., Kořínek, R., and Žůrek P. (2006) Cultural Monument Jeroným Mine. Czech Republic – Contribution to the Geomechanical Stability Assessment. Publs. Inst. Geophys. Pol. Acad. Sc., M-29(935), Warszawa, 137–146.

Kaláb, Z., Knejzlík, J., Kořínek, R., Kukutsch, R., Lednická, M., and Žůrek, P. (2008a) Contribution to Experimental Geomechanical and Seismological Measurements in the Jeroným Mine. Acta Geodynamica et Geomaterialia, Vol. 5, No. 2(150), 213–223.

Kaláb, Z., Hrubešová, E., Knejzlík, J., Kořínek, R., Kukutsch, R., Lednická, M. and Žůrek, P. (2008b) Mine Water Movement in Shallow Medieval Mine Jeroným (Czech Republic). In: Rapantová, N. and Hrkal, Z. (Eds): Mine Water and the Environment. Proceedings of 10th International Mine Water Association Congress. Karlovy Vary. VŠB-Technical University of Ostrava; full paper at CD, No. 37, 11 pages.

Kaláb, Z. and Lednická, M. (2009) Potential Representation of Increasing of Mine Water Level on Middle Horizon of the Jeroným Mine. Transactions, No. 2, in print.

Kang, S. S. (2000) Measurement and Interpretation on Stress History of a Limestone Deposit. A dissertation for Degree of Doctor of Philosophy, Kumamoto University.

Knejzlík, J. and Rambouský, Z. (2008) Recent Solution of the Distributed Control and Measurement System in the Jeronym Mine - Modular System. Acta Geodynamica et Geomaterialia, Vol. 5, No. 2(150), 205–212.

Knejzlík, J., Rambouský, Z., Souček, K., and Staš, L. (2008) Second Generation of Conical Strain Gauge Probe for Stress Measurement in Rock Massif. Acta Geodynamica et Geomaterialia, Vol.5, No. 3 (151), 1–9.

Nakamura, N., Ohkubo, R., Obara, Y., Kang, S. S., Sugawara, K., and Kaneko, K. (1999) Rock Stress Measurement for Limestone Open Pit Mine. In: Proc. of 5th Int. Symp. on Field Measurements in Geomechanics, Singapore, Balkema, Rotterdam, 375-380.

Staš, L., Knejzlík, J., and Rambouský, Z. (2005) Conical Strain Gauge Probe for Stress Measurements. EUROCK 2005 – Impact of Human Activity on the Geological Environment – Konečný (ed). Balkema, London, 587–592

Staš, L., Souček, K., and Knejzlík, J. (2007) Conical Borehole Strain Gauge Probe Applied to Induced Rock Stress Changes Measurement. 12th International Congress on Energy and Mineral Resources. Proceedings, Oviedo: Consejo superior de celegios de ingenieros de minias, 507–516.

Žůrek, P., and Kořínek, R. (2004) Opening of the Medieval Jeroným Mine in the Czech Republic to the Public. Acta Montanistica Slovaca, Ser. 8., 96–100.

Chapter 7
Seismic and Geodetic Observations
of Subsidence Trough Development Over
a Longwall Face in a Coal Bed Under Extraction

Michał Bogusz and Maciej Mendecki

Abstract The rock layers located in the vicinity of mining works are deformed as a result of mining exploitation. Such deformations cause a development of subsidence trough over a coal bed under extraction. The distortions also contribute to the appearance of seismic events in the mining works area. The goal of measurements was to find a relation between these phenomena. The research was carried out over longwall face 183, coal bed 418, belonging to the coal mine "Pokój" in Ruda Śląska. The size of surface deformation was determined from GPS measurements performed on the benchmarks (BM) located over the longwall face under study. The GPS data were also used to determine the maximum surface subsidence velocity (MSSV) in the trough area and to assess the approximate time of occurrence of maximum surface subsidence velocity (MSSV average time). Finally, the relations between subsidence parameters and seismic activity were sought.

7.1 Introduction

Mining exploitation is indispensably connected with the appearance of tremors. The mining sensors record low-energy mining tremors and higher energy tremors. The former are related to coal exploitation and the latter are caused by interaction between exploitation stress and regional geology or tectonic stresses (Gibowicz 1989; Idziak et al. 1999; Stec and Błaszczyk 2008). Apart from tremors, the gradual

M. Bogusz
University of Silesia, Faculty of Earth Sciences, ul. Będzińska 60, 41-200 Sosnowiec, Poland
and
Kompania Weglowa S.A, Coal Mine "Bobrek-Centrum", ul. Konstytucji 76, 41-905 Bytom, Poland
e-mail: m.bogusz@kwsa.pl

M. Mendecki (✉)
University of Silesia, Faculty of Earth Sciences, ul. Będzińska 60, 41-200 Sosnowiec, Poland
e-mail: mmendecki@us.edu.pl

A.F. Idziak and R. Dubiel (eds.), *Geophysics in Mining and Environmental Protection*, Geoplanet: Earth and Planetary Sciences 2, DOI 10.1007/978-3-642-19097-1_7, © Springer-Verlag Berlin Heidelberg 2011

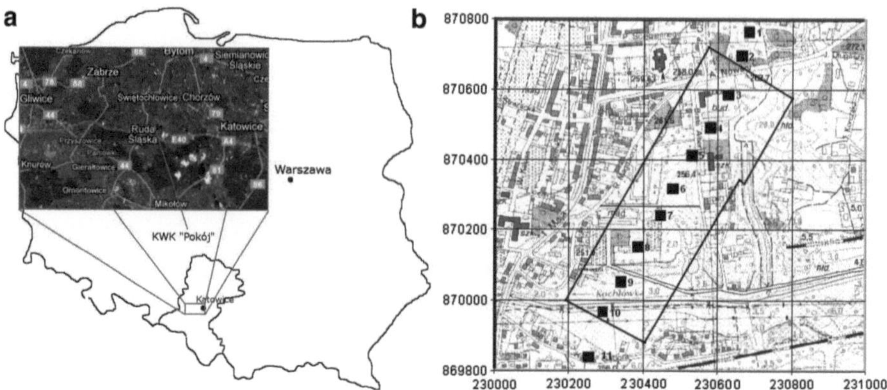

Fig. 7.1 Location of the research area (**a**) and survey profile with longwall contour in the background of topographic map (**b**)

deformation of the ground surface is observed as a result of the influence of exploitation on rock mass. A rock translocation of roof rocks occurring during the coal bed exploitation can be transferred onto the surface, causing the development of subsidence trough (Knothe 1980; Szpetkowski 1980; Kowalski 1997).

Surface deformations are monitored by geodetic methods. The most popular devices used in geodetic measurements are the dumpy levels and total stations. Recently, however, the ground displacements are often being determined with the use of GPS techniques, because they are faster and more precise (Góral and Szewczyk 2004; Białek et al. 2007).

The goal of the research presented in this paper is an attempt to link the observed surface deformation measured with GPS equipment with the seismicity induced by longwall face 183, coal bed 418. The studied longwall face belongs to coal mine "Pokój" which is located in Ruda Śląska, Poland (Fig. 7.1a).

7.2   Research Description

The investigated longwall face and the coal bed belong to Ruda stratum. The Upper Carboniferous deposit in the "Pokój" coal mine area is covered by the Quaternary sediments with a thickness around 100 m. The shale with light gray sandstones and also coal shale or sandy shale (borehole card "Szyb Wanda") occur in the coal bed overburden.

The longwall face 183 was extracted from 21 May 2007 to 28 November 2008. The coal bed was at an average depth of about 700 m and a parallel-to-strike system of longwall extraction with caving was applied. In the research area, beds dip to the SSW at a small angle (several degrees). The exploitation was led in the direction from SSW to NNE. The longwall face 183 neighbors the Second West Main Fault on the ESE and the gob of longwall face 182 on the WNW.

Observations of benchmarks (BMs) positions were carried out during the period from August 2007 to October 2008 (14 measurement cycles) with the GPS equipment of Trimble. Time intervals between measurement cycles amounted to about 1 month, but the last interval was longer (4 months).

The survey line consists of 11 benchmarks (Fig. 7.1b). The measurement points were situated along a straight line, at about 100 m distances. The survey profile runs along the longwall face 183. During the measurement period, some points were lost. The last measurement on BM 4 was carried out on 1 February 2008, because construction works have started near this point afterwards. BMs 6 and 7 existed until 27 June 2008. After this measurement cycle, the area around these points was flooded as a result of subsidence motion.

The collected data were post-processed with Trimble Geomatics Office software (TGO).

7.3   Results: Seismic Activity

Seismological observations were ran by the mining geophysical station. 726 tremors in the energy range of 10^3–10^6 J were analyzed. Mining data were used to make a map of seismic event positions. Figure 7.2 shows tremors location in the

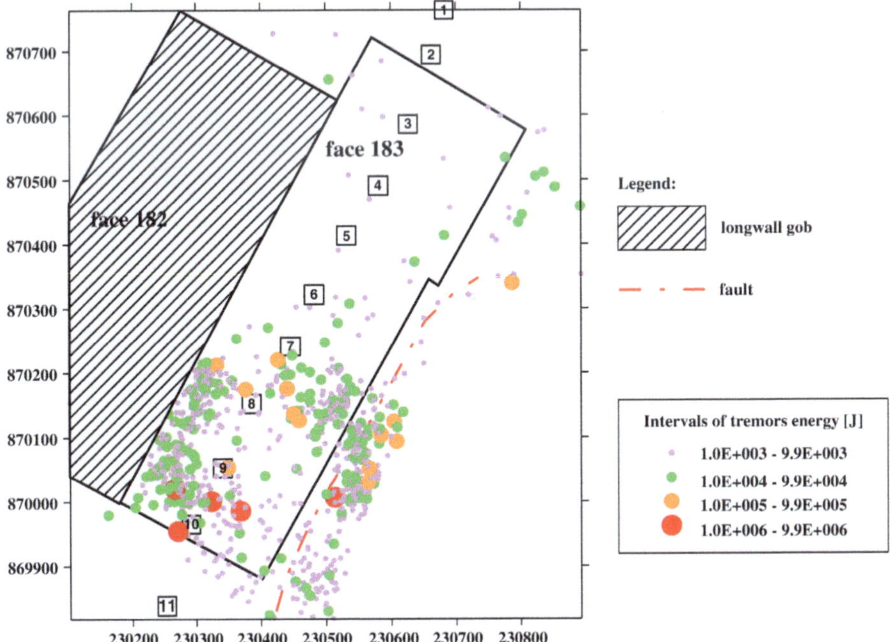

Fig. 7.2   Seismic event position map (coordinates in '65 array, zone V)

background of the longwall faces, the geological structure and the benchmarks installed over the studied longwall face 183.

The positions of tremors show that the hypocenters are distributed unevenly. There are areas with high concentration of earthquakes. The first such area is situated in the initial part of the longwall face. Many low-energy tremors but also high energy ones (magnitude of the order of 10^6 J) occur there. Another increased activity zone is located in the place where the longwall face 183 borders the exhaust longwall face 182. The third zone has been found along the fault. The last area of increased activity is situated underneath BMs 7 and 8. Probably this concentration is caused by the presence of a gob over the studied longwall face.

7.4   Results: Geodetic Observations

The results of geodetic observations are presented as a comprehensive plot (Fig. 7.3). It shows the development of subsidence trough along the survey profile in the measurement cycles. In the bottom part of Fig. 7.3 there is also shown a scheme of presumable longwall face locations at the moment of measurement.

The monotonous subsidence was observed in initial measurement cycles (14 September to 30 November 2007) along the survey profile. In the next cycle (14 December 2007), the moment when the trough bottom started to develop around benchmark 9 is visible. In next stages, the trough minimum deepens and moves horizontally according to the longwall face advance direction. This process was

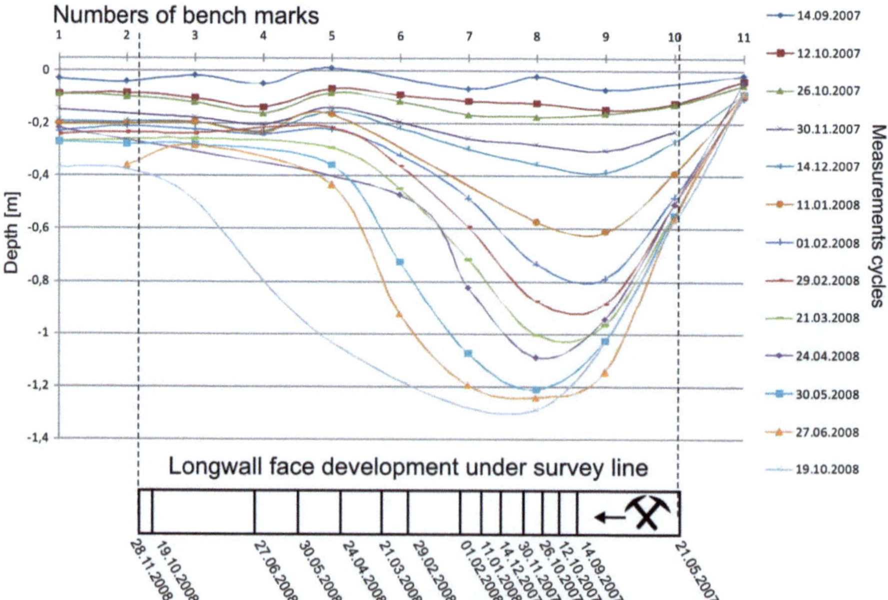

Fig. 7.3 Changes of surface subsidence measured on benchmarks

finished when the bottom of the basin stopped developing in the area of benchmark 8 (May–June 2008). In the last cycles, the vertical development of trough beneath the rest of points was observed.

7.5 Determination of Average Time of Maximum Surface Subsidence Velocity (MSSV Average Time)

The collected data were plotted as a subsidence against time. We wanted to adapt an analytical function to measurement data in order to determine the average time of the maximum surface subsidence velocity (MSSV) on benchmarks. The adapted Fermi function (Skrzypczak and Szefliński 1995; Sawieliew 1998) was used as a first good approximation of subsidence model in the following form:

$$h(t) = a_1 \left[1 - \left(e^{\frac{t-a_2}{a_3}} + 1 \right)^{-1} \right] \tag{7.1}$$

where $h(t)$ is the subsidence in meters, t is time in days, a_1 is the maximum subsidence with time tending to infinity, a_2 is the MSSV average time, and a_3 is the curvature factor.

The adapted Fermi function is similar in shape to functions applied in mining geodesy to describe surface subsidence with time (e.g., Knothe 1980; Kochmański 1980):

$$w(t) = \sum_{i=1}^{n} w_{ki}(1 - e^{-ct_i}), \tag{7.2}$$

where w_{ki} is the final subsidence value in ith field, $t_i = t' - t_{ei}$, while t' means the measurement time and t_{ei} is the average time of exploitation for ith field counted since time $t_i = 0$.

Parameter a_1 corresponds to the final subsidence w_k in infinite time. When we compared the adapted function (1) with Kochmański's function (2), we noticed that the parameter a_3 may be related to the inverse parameter c which characterizes velocity of ground motion caused by mining exploitation.

In the adapted function, parameter a_2 is, by definition, the MSSV average time, which results from the properties of the Fermi function. In Kochmański's function, however, the time t_i is a difference between the measurement time and the average time of the ith field exploitation. There is also another difference between the two functions. In the equation proposed by T. Kochmański, the whole coal deposit is considered as composed of smaller fields exploited at different average times. In the case of Fermi function it was noticed that this function is applicable only in the case the basin has been fully developed. Therefore, the approximation was made only for BMs 7, 8, 9, and 10.

It is easy to determine the surface subsidence velocity as a derivative with respect to time:

$$v(t) = \frac{dh(t)}{dt} = \frac{a_1}{a_3} \exp\left(\frac{t - a_2}{a_3}\right) \cdot \left(\exp\left(\frac{t - a_2}{a_3}\right) + 1\right)^{-2}. \qquad (7.3)$$

Assuming that $t = t_{ave} = a_2$ for the maximum surface subsidence velocity, (7.3) simplifies to the following form:

$$v_{\max} = \frac{a_1}{a_3} \cdot 0.25. \qquad (7.4)$$

Plots in Fig. 7.4 show the distribution of the measurement data with errors, the fitted adopted Fermi function (solid line) and the velocity distribution (dashed line). Plots also include the computed values of the following parameters: MSSV average times and MSSVs.

The least squares method was applied to fit the parameters of the adapted function to measured data. Test χ^2 was used to verify the conformity. The test proved that for BMs 7, 9 and 10 there was no reason to confirm the hypothesis about wrong selection of approximated function to measured points. For BM 8, however, the hypothesis about wrong selection of approximated function to the measured points was confirmed, although the statistical significance for this solution is close

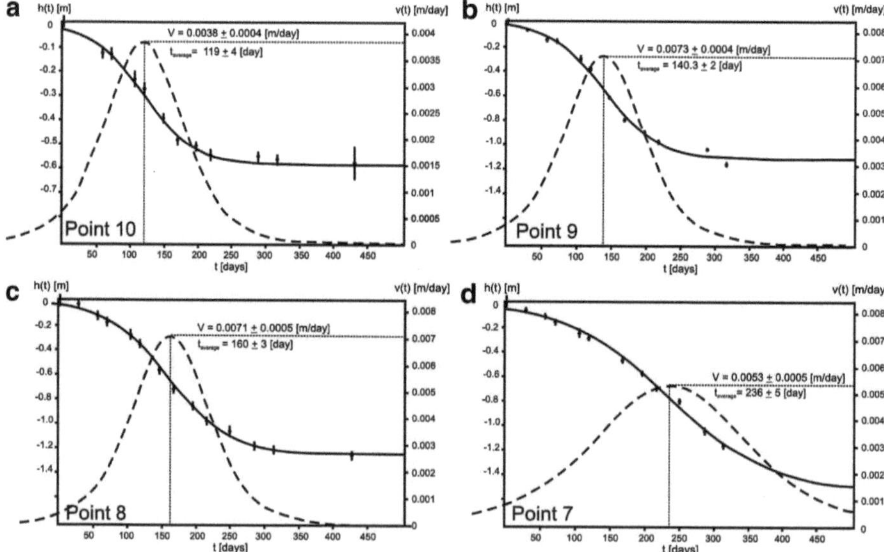

Fig. 7.4 Plots of surface subsidence (solid line) and subsidence velocity (dashed line) against time for benchmarks: a) BM No. 10, b) BM No. 9, c) BM No. 8 and d) BM No. 7

to statistical significance limit 0.05. Hence, we assume that approximation for BM 8 is acceptable.

The calculated values of parameter a_2 and maximum velocity values are compared in Table 7.1. The table provides also information about total longwall face advance, the time of longwall face location under benchmarks, and the difference between these two events.

In addition to MSSV and MSSV average times, we also calculated the values of parameter a_3. Table 7.2 shows the values of a_3 for particular benchmarks and the values of factor c. As shown by Szpetkowski (1980), the value of factor c for the Upper Silesia Coal Basin ranges from $c = 0.5$/year (for deformations in overlaying thick and solid sandstones) to $c = 5$/year and more (for deformations in overlying brittle rocks or plastic rock).

Errors of parameter c were calculated by applying the total differential in the following form:

$$\Delta c = \left| \frac{\partial c}{\partial a_3} \right| \cdot \Delta a_3 = \frac{\Delta a_3}{a_3^2}. \tag{7.5}$$

The values of parameter c calculated for consecutive benchmarks show that the overburden contains brittle rocks or plastic rock. The c value decreases and for BM 7 it achieves the limiting value (near 5/year). This is confirmed by the information from borehole profile which documented the presence of shale, sandy shale and coal shale in the overburden.

Table 7.1 Calculated parameters of trough development with longwall face advance

BM No.	Total longwall face advance (m)	Time of longwall face location under BM	MSSV average time ($t_{ave} = a_2$)	Time difference (month)	Maximum surface subsidence velocity (mm/day)
10	29	Middle of June 2007	Middle of December 2007	6	3.8 ± 0.4
9	126	End of August 2007	Beginning of January 2008	4	7.3 ± 0.4
8	227	Middle of November 2007	Middle of January 2008	2	7.1 ± 0.5
7	341	Middle of January 2008	Beginning of April 2008	2.5	5.3 ± 0.5

Table 7.2 Calculated values of a_3 and c with errors

BM No.	Values of a_3 (day)	Values of a_3 (year)	Values of factor $c = 1/a_3$ (1/year)
10	36.3 ± 3.6	0.0995 ± 0.0099	10.1 ± 1.0
9	35.3 ± 1.8	0.0967 ± 0.0049	10.3 ± 0.5
8	44.6 ± 2.9	0.1221 ± 0.0079	8.2 ± 0.5
7	74.1 ± 5.2	0.203 ± 0.0142	4.9 ± 0.3

7.6 Relations Between the Observed Surface Deformation and Seismicity Induced by Coal Mining

The beginning of longwall face 183 was a place where we observed the longest time difference between the appearance of maximum surface subsidence velocity and the time when the longwall face is located under BMs. Figure 7.2 shows the area with high seismic activity, with energies of 10^6 J. When the longwall face was advancing, the time differences were decreasing and the seismic activity was decreasing too. Probably these relations might be a result of the fact that in the initial stage of coal mining the rock mass was not disturbed and the trough development pace was very slow, so that a large stress was present in the rock mass due to the pressure of overburden; this was manifested as a large number of weak tremors as well as those of higher energy. When coal mining advanced, the rock mass was disturbed and the trough development was faster, so that a smaller stress was present in the rock mass, which reduced the seismic activity and shortened the above-mentioned time delay.

7.7 Conclusions

1. The maximum level of seismic activity is observed at the beginning of exploitation.
2. The concentration of seismic activity is located in the vicinity of gobs and the fault.
3. The seismic activity area moves together with the longwall face advance.
4. It was demonstrated that GPS measurements can be used for deformation monitoring, including the subsidence trough development.
5. The distance between the surface point and the beginning of the exploited longwall face has an influence on the time of appearance of the maximum surface subsidence velocity on this point. In our study, the time difference was initially 6 months, then 4 months and eventually fell to 2 months.

References

Białek J, Sokoła-Szewioła V., and Opałka K. (2007) *Prędkość obniżeń punktów powierzchni terenu a aktywnosc sejsmiczna gortworu* [Surface subsidence rate and seismic activity of the rock mass], Zeszyty Naukowe Politechniki Śląskiej Seria Górnictwo z. 278 (in Polish).

Gibowicz S. (1989) *Mechanizm ognisk wstrząsów górniczych* [Focal mechanism of mining tremors], Publs. Inst. Geophys. Pol. Acad. Sc. M-13(221) (in Polish)

Góral W., and Szewczyk J. (2004) *Zastosowanie technologii GPS w precyzyjnych pomiarach deformacji* [GPS technology in precise measurements of deformations], Uczelniane Wydawnictwo Naukowo–Dydaktyczne AGH, Kraków (in Polish).

Idziak A., Teper L., and Zuberek W. M. (1999) Sejsmiczność a tektonika Górnośląskiego Zagłębia Węglowego [Seismicity and tectonics of the Upper Silesian Coal Basin], Wydawnictwo UŚ, Katowice (in Polish).

Knothe S. (1980) *Wpływ czasu na przebieg przemieszczeń, deformacji górotworu i powierzchni terenu wywołanych eksploatacja górniczą* [Effect of time on the behavior of mining-induced deformation of the rock body and surface] In: "Ochrona powierzchni przed szkodami górniczymi", M. Borecki, ed., Wyd. "Śląsk", Katowice (in Polish).

Kochmański T. (1980) *Teoria Tadeusza Kochmańskiego* [Tadeusz Kochmański theory], In: "Ochrona powierzchni przed szkodami górniczymi" M. Borecki, ed., Wyd. "Śląsk", Katowice (in Polish).

Kowalski A. (1997) *Uwarunkowania deformacji powierzchni* [Surface deformations], In: "Ochrona obiektów budowlanych na terenach górniczych" J. Kwiatek, ed., Wyd. Głównego Instytutu Górnictwa, Katowice (in Polish).

Sawieliew I.W. (1998), *Wykłady z fizyki*, vol. 3 [Lectures in physics], Wydawnictwa Naukowe PWN, Warszawa, Third Edition (in Polish).

Skrzypczak E., and Szefliński Z. (1995) *Wstęp do fizyki jądra atomowego i cząstek elementarnych – wykłady* [Introduction to nuclear physics – lectures], Wydawnictwo Naukowe PWN, Warszawa (in Polish).

Stec K., and Błaszczyk E. (2008), *Charakterystyka procesów zachodzących w ogniskach wysokoeergetycznych wstrząsów występujących w czasie eksploatacji ściany 17 w pokładzie 361 w KWK Knurów* [Processes in the foci of high-energy shocks during mining of longwall 361 in the Knurow colliery], Gospodarka Surowcami Mineralnymi, 24, 2/3 (in Polish).

Szpetkowski S. (1980) *Charakterystyka wpływów robót górniczych na górotwór i na powierzchnię terenu* [Characteristics of the effect of mining works on the rock mass and surface], In: "Ochrona powierzchni przed szkodami górniczymi", M. Borecki, ed., Wyd. "Śląsk", Katowice (in Polish).

Chapter 8
Suggestion of Complex Monitoring of Undermined Territory

Hana Doležalová

Abstract Underground mining of coal deposits in Ostrava-Karvina region causes movements and deformations on the surface where a subsidence depression starts to develop. Real displays of undermining can be only found out by repeated in situ measurements; geodetic and geophysical methods can be used. Photogrammetry enables to observe the whole subsidence depression and to create digital terrain model. GPS is used to monitor both vertical and horizontal surface changes. Levelling is used to find out even small changes, but only in vertical subsidence. Therefore, GPS is generally used to monitor the subsidence depression in its whole extent, while levelling is used to specify small height changes near the margin of the subsidence depression. Geophysical methods may contribute to the monitoring of the undermined territory and they may enable to understand the processes induced by underground mining in the rock massif.

8.1 Introduction

Due to underground exploitation of coal deposits, there occur vacant spaces in Ostrava-Karviná region. The overlaying strata can then sag or fall into these spaces. This transport of material gradually passes through the overburden and it results in movements and deformations on the surface where a subsidence depression starts to create. The subsidence of the undermined territory is often modelled but the real surface displays of undermining can only be found out by in situ measurement. The most common way to observe the surface displays of undermining is the geodetic height surveying, mostly by levelling. But levelling only defines height changes on the surface, not horizontal shifts. These can be gained from repeated surveying of stabilized points by GPS method. The method of aerial photogrammetry makes it

H. Doležalová (✉)
Institute of Geonics, Academy of Sciences of the Czech Republic, 708 00 Ostrava-Poruba, Czech Republic
e-mail: dolezalova@ugn.cas.cz

A.F. Idziak and R. Dubiel (eds.), *Geophysics in Mining and Environmental Protection*,
Geoplanet: Earth and Planetary Sciences 2, DOI 10.1007/978-3-642-19097-1_8,
© Springer-Verlag Berlin Heidelberg 2011

possible to create a digital terrain model. Each of these geodetic methods has its pro and con, and the choice of optimal surveying method for the undermined territory depends mainly on the purpose of surveying and on actual conditions in a given locality. Generally, geodetic methods provide information on surface changes on the undermined territory, and to deduce from them the changes of the rock massif due to underground exploitation may not be easy. For this purpose, geophysical methods can be used. They can provide information on changes of the rock massif caused by exploitation.

8.2   Geodetic Methods on Undermined Territory

Geodetic methods are the most common way to observe the effects of undermining on the surface. The proper method can provide site, height or spatial state of the terrain with adequate accuracy. While we find out the state of surface at a certain time by single geodetic surveying, it is possible to observe the changes of the surface and the process of creation of the subsidence depression by repeated geodetic surveying.

The real movements and deformations of the surface can be found out either directly, by surveying, or indirectly, by computing from other measured values. The choice of geodetic method depends primarily on the purpose of observation and also on local conditions. It is necessary to choose more accurate methods in localities where smaller movements are expected than in localities where sizable surface changes are expected. Local conditions can also complicate or forbid the usage of some methods. According to the purpose of observation, methods of height or site surveying or methods of spatial position surveying of points are chosen.

Observation networks as sets of fixed points on objects or on surface are built for observation of mining effects on the surface. These networks enable repeated surveying of stabilized points and evaluation of movements and deformations progress in time. Building of an observation network (type of stabilization and layout of points) depends mostly on its purpose and on the size of supposed subsidence depression. The method of surveying and local conditions have to be taken into account too.

Beside geodetic methods that are surveying from one point to another, it is also possible to use a method that enables to observe the entire subsidence depression – method of aerial photogrammetry. The principle of photogrammetry rests in taking of photographs of the surface or its objects and in subsequent elaboration of overlap photographs. The accuracy decreases with the distance from the observed object and it is ca. first tens of centimetres. That is why the usage of this method for observation of changes in the subsidence depression is recommended mainly in localities where sizable changes are expected.

The aerial photogrammetry was chosen to observe the undermined surface of the whole subsidence depression that was creating near the city of Karviná, above two exploited mining panels. The aerial photogrammetry was done repeatedly once a year to gain repeated surveying from which the changes of the subsidence depression could be seen. As there was also GPS surveying done repeatedly, we could find out the accuracy of the photogrammetry by comparing the coordinates of points surveyed by GPS that were also signalised during aerial photogrammetry (red points in Fig. 8.1). The coordinates of most of these points, gained from both GPS and photogrammetry, did not differ by more than 10 cm, rarely there were points with coordinates differing by more, ca. 30 cm at most.

The method of aerial photogrammetry markedly enriched and complemented the GPS surveying because the observation network cannot be made of regular net of fixed points in all parts of area of concern, so the model done from GPS surveying would be distorted in parts of small density of points. The aerial photogrammetry provided data for the digital terrain model of the observed subsidence depression and also its surroundings (out of more than 17,000 points, Fig. 8.1). Repeating aerial photogrammetry in an interval of 1 year then showed on places of major surface changes caused by undermining. However, in consideration of a lower accuracy of this method, it is not optimal to be used for observation of small changes, for example near the margins of subsidence depression.

GPS is a satellite system to determinate spatial position of points. Using GPS method to observe the changes on the surface of the undermined territory is very suitable because it provides a possibility of finding out the spatial position of a point in a short time with the accuracy that is generally higher than that of aerial photogrammetry. With repeated surveying, not only subsidence but also horizontal shifts of points, or the vector of the overall movement can be observed. In areas where fast changes happen in the overburden of exploited deposit, GPS can effectively solve the problem of geodetic connection of undermined points to stable points that are out of reach of undermining, even several tens of kilometres.

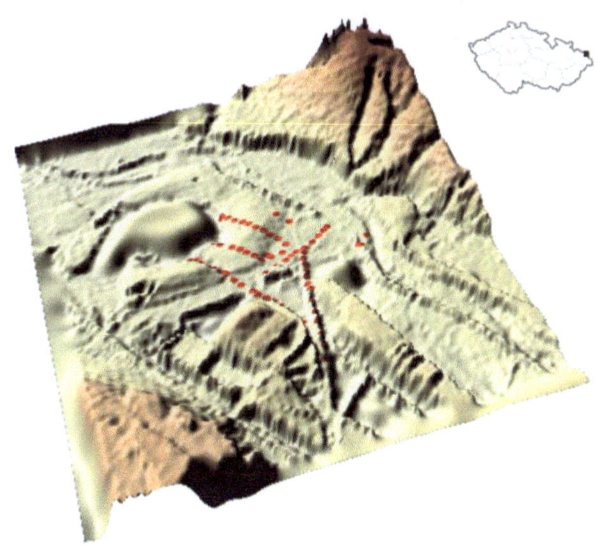

Fig. 8.1 Digital terrain model from aerial photogrammetry (Doležalová et al. 2008)

There was a systematic geodetic monitoring done by GPS method on the observation network near the city of Karviná from 2007 to 2008. It has shown on both advantages and disadvantages of using this method for the observation of undermining effects on the surface. Single points of the observation network were surveyed by static method with observation at least 10 min per point. Thanks to stabilization of points of the observation network in the form of lines and scattered points, it is possible to express not only subsidence of single points, but also subsidence in profiles and in area. But the areal evaluation is unfavourably affected by irregularity of the point network, because due to the reclamation, the points could not be stabilized in all parts of the territory affected by exploitation. With regard to surveying the spatial position of points, not only vertical subsidence but also horizontal shifts, and/or deformations could be found out from the GPS surveying.

Both point and profile evaluation enabled the observation of development of subsidence and it has shown on places of irregular development of subsidence depression. The subsidence of points in one of the profiles is shown in Fig. 8.2. It is apparent that dramatic changes happen on the section between points 14 and 18; the distance is ca. 200 m, difference in subsidence of these points is almost 80 cm. The cause may be seen in complicated tectonic situation, together with earlier exploitation in the overburden in this part of the observed area.

Subsequent evaluation of horizontal shifts revealed that several end points of the profiles are not oriented to the centre of observed subsidence depression, as one would suppose, but they are oriented elsewhere. Directions of measured horizontal movements of some of the monitored points in the period 2007–2008 are shown in

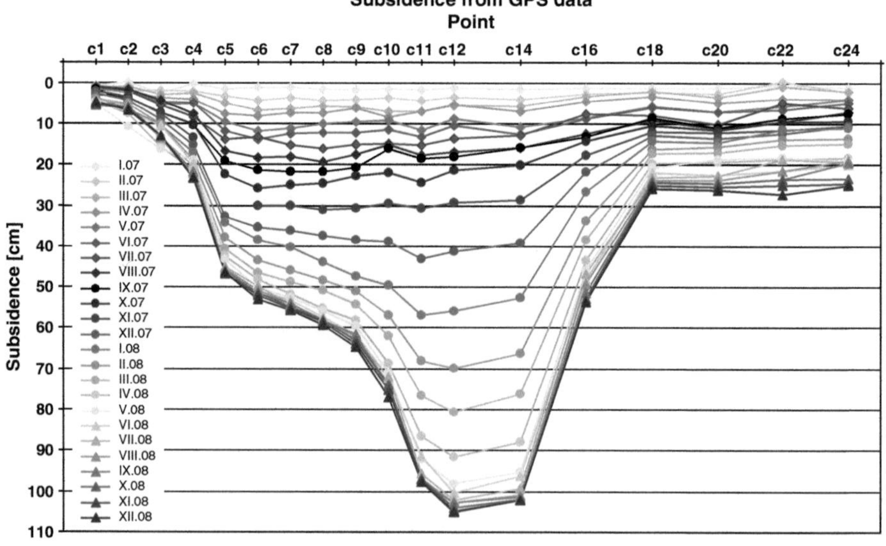

Fig. 8.2 Subsidence gained from GPS surveying (Doležalová et al. 2009)

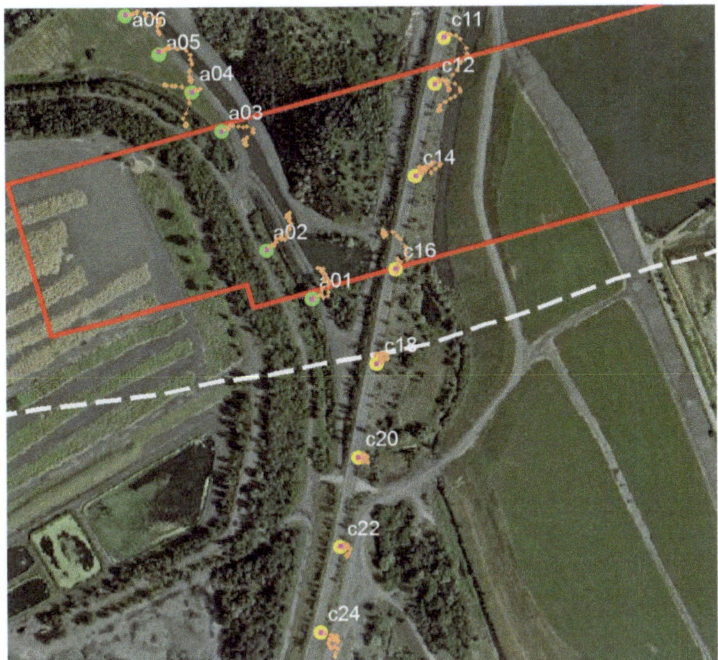

Fig. 8.3   Horizontal movements of points monitored by GPS in the period 2007–2008 (Doležalová 2009)

Fig. 8.3; curves are oversized, maximum change of stabilized point position was 23 cm. Thanks to such evaluation of direction of horizontal shifts, it could be stated that these end points do not change their position due to exploitation of mining panels under the centre of observed subsidence depression, but they are affected by exploitation in the surroundings. This fact would not be known if subsidence was evaluated only. The advantage of GPS method is the possibility to observe not only subsidence but also size and direction of horizontal shifts. Such evaluation is markedly contributing to the understanding of processes on the surface of the undermined territory.

A disadvantage of GPS method can be seen in its accuracy limitation for observation of small movements. A temporary upheaval of some points was detected by GPS method on the observation network near Karviná. However, the upheaval was always rather random and no marked trend developed there. In addition, the upheaval was only up to 2 cm and this may not be a uniquely determined change, considering the total accuracy of GPS surveying. So the disadvantage of GPS method may be its accuracy that is generally lower in height than in horizontal position. But it is possible to improve the accuracy of GPS results by choosing another surveying procedure with longer static observation. However, such surveying is more time and financially consuming and it is rarely used to observe the effects of undermining on the surface. More often, levelling is used for more accurate height surveying.

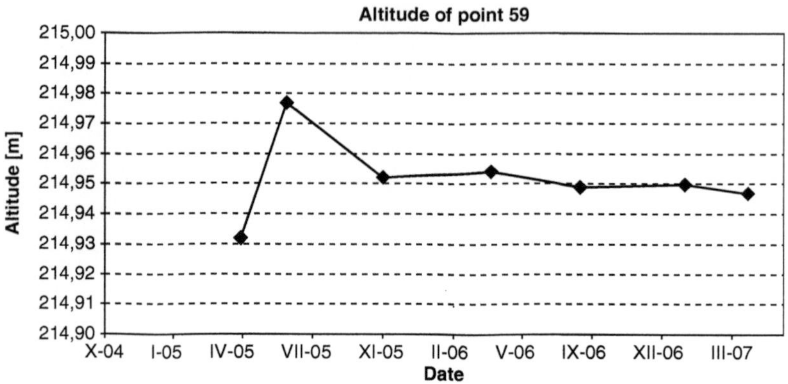

Fig. 8.4 Altitude changes of levelled point

The levelling is most widely used and for common practice it is the most exact
type of height surveying. Thanks to its accuracy, levelling is optimal method for
recognition of subsidence on the undermined territory and probably the most
optimal method for observation of small height changes near the margin of subsidence depression. Thanks to many advantages, geometric levelling from the centre
is mostly used. Accurate or technical levelling is used for determination of points'
altitudes in the surface point mine network in the Czech Republic, according to
regulation of the Czech mining office board (ČBÚ No. 435/1992 Sb.). Both types of
levelling make it possible to observe height changes on the surface of the
undermined territory with higher accuracy than is usually offered by GPS. It is
possible to observe even small subsidence and perhaps even an upheaval that may
also happen near the margins of subsidence depression.

The levelling that was repeatedly done at the margin of subsidence depression
near the municipality Dětmarovice detected (with sufficient accuracy) changes of
altitude of observed points. Not only subsidence but also upheaval was recorded
(see Fig. 8.4). These height changes were caused by redistribution of stress in the
rock massif due to underground mining, by changes of groundwater regime and
further by stabilization of observed points near the water flow. Other geodetic
methods would hardly detect such small height changes with sufficient accuracy.

A disadvantage of levelling is providing data only on height state of the surface
so it is not possible to observe the direction of the overall movement of points but
only its vertical component.

8.3   Geophysical Methods on Undermined Territory

While geodetic methods observe surface changes caused by underground exploitation, geophysical methods may clear up the processes that happen inside the rock
massif. The advantage of geophysical methods may be also seen in the fact that they

can be used for observation of rock massif deformations already in the stage when no geodetically detectable effects have already appeared on the surface.

The usage of geophysical methods in the subsidence depression is conditional on creation of physical-mechanical changes in the rock massif. During transformation of the rock massif due to underground mining, cracks create, rock strength decreases, etc. Repeated measurement by geophysical methods makes it possible to observe the progress of the rock massif loosening in consequence of underground exploitation.

From the surface geophysical methods, gravimetric measurement can be used, because density decreases with growing porosity in the loosened parts of the rock massif. Also seismic measurement can be used, because also velocity of longitudinal waves propagation decreases with growing porosity. Transformation of the rock massif also displays with changes in resistivity, which can be observed by geo-electrical measurement. Differently loosened parts of rock massif can be also distinguished by heat conductivity of rocks and penetration depth of temperature changes by using geothermal measurement of the course of temperature field. Also radiometric measurement can be used, because pores and cracks induced in loosening massif cause increased radioactive emanations (Müller et al. 1985).

Beside surface geophysical measurement, we can also use a complex of logging methods in boreholes to locate crack zones. The loosened zones created due to exploitation can be located by radioactive, electric or acoustic logging (Müller et al. 1985; Mareš et al. 1990).

Geophysical survey done by firm Geotest Brno, a.s., on several localities near the margins of subsidence depression in Karviná region showed that not only individual methods separately, but above all the complex evaluation of several methods together is able to refer to changes occurring in the rock massif due to underground mining. Summary interpretation of repeated logging methods (neutron–neutron logging, gamma–gamma logging and acoustic logging) showed deformations of the rock massif at a depth of 27–40 m. These changes of stress-strain state give evidence about deformation of the rock massif in consequence of undermining. Also repeated seismic tomography in the borehole–surface system revealed this zone of failure. Repeated surface survey by symmetrical resistivity profiling detected an existence of resistivity anomaly that also shows on changes in the rock massif. These changes were also confirmed by precise inclinometry which is rather geo-technical method of monitoring (Bláha et al. 2007).

8.4 Suggestion of Complex Monitoring of Undermined Territory

Each of the geodetic and geophysical methods can provide information on changes in the subsidence depression. However, to understand the progress of the whole subsidence depression, a complex monitoring is needed, i.e., monitoring by a complex of geodetic, geophysical and perhaps even geo-technical methods, on the basis of observation of changes both on the surface of undermined territory and in the rock massif stricken by underground mining. Individual methods provide data on height, site or spatial position changes of observed points, eventually the whole surface of the undermined territory, and also data on changes of physical parameters of the rock massif. Complex analysis of the results of individual survey methods enables then complex evaluation and understanding of processes that happen due to underground mining.

Complex monitoring of the subsidence depression should enable to observe changes in its whole extent, perhaps even in its vicinity. It should be done on the basis of repeated measurements with different methods to find out (with sufficient accuracy) the changes on the surface and in the rock massif. After an optimal stabilization of points of the observation network, it is suitable to use the GPS method to observe changes of the spatial position of points. If the points are stabilized roughly in a regular net, the development of surface changes in the range of the whole subsidence depression can be observed on the basis of surface models. From the three-dimensional coordinates of points gained from GPS surveying, not only vertical subsidence but also horizontal shifts of points can be observed. Observing the directions of these shifts is markedly contributing to better understanding of the overall movement progress on the surface of the undermined territory. The common usage of GPS method on the undermined territory may not be sufficient for observation of small surface changes with satisfactory accuracy. That is why it is necessary to observe the areas with small changes also with other, more accurate method. Levelling is able to detect height changes even in the range of several millimetres and so it can properly complete the GPS surveying, for example in margin parts of the subsidence depression or generally in those parts of the undermined territory where only small surface changes happen. Monitoring by complex of geophysical and maybe even geo-technical methods is sufficient to be done too in areas where levelling could not uniquely determine the margin the subsidence depression or in areas of higher importance (objects on the undermined territory). It enables to observe the changes of the rock massif in areas where the surface effects of underground mining do not display yet or they are too small to be recorded by geodetic surveying. Geophysical methods used for measurements on the surface and in the boreholes can detect changes of physical and/or physical-mechanical properties. From them, the processes in the rock massif can be deduced.

Recultivation of landscape is often in the process on the undermined territory and the surface is partly covered with mine spoils. That is why it is suitable to complete the observation of subsidence depression with a method of indirect data capture, because it is not possible to stabilize points and repeatedly survey them by some common geodetic method in the areas of active recultivation. So, complex monitoring should include also detection of overall shape of the subsidence depression in the form of digital terrain model by method of aerial photogrammetry. In some cases, this method can be replaced by a special application of radar interferometry (e.g., permanent scatterers method) that can be an optimal method to observe surface changes in the range of the whole subsidence depression, especially when it is evaluated together with land geodetic surveying.

8.5 Conclusion

Geodetic monitoring provides concededly the most complete information on subsidence and shifts of the surface in the areas of subsidence depressions that are created due to exploitation of coal deposits in the Karviná region. Geodetic monitoring should be complemented with geophysical monitoring to evaluate also the processes in the rock massif. Complex geodetic and geophysical monitoring of the undermined territory based on using methods of aerial photogrammetry (or radar interferometry), GPS and levelling and also geophysical methods on the surface and in the borehole would provide not only information on changes of the surface of undermined territory, but also information on changes happening under the surface of the subsidence depression, directly in the rock massif, and so it would enable the complex evaluation of processes induced by underground mining.

Acknowledgement This paper has been prepared in the frame of grant projects of Czech Science Foundation, No. 105/07/0878 and No. 105/07/1586.

References

Bláha, P., Doležalová, H., Duras, R., Lukeš, J., and Müller, K. (2007) Rock Massif Changes Caused by Undermining. Transactions of the VSB-Technical University of Ostrava, Civil Engineering Series. VII, č.2/2007, p. 17-30. ISSN 1213-1962 (in Czech).

Doležalová, H. (2009) Analysis of Geodetic and Geophysical Monitoring of the Undermined Area. Ph.D. thesis. Ostrava: ÚGN AV ČR VŠB-TU Ostrava, 2009. 127 p. (in Czech).

Doležalová, H., Kajzar, V., Souček, K., and Staš, L. (2009) Evaluation of Mining Subsidence using GPS data. Acta geodynamica et geomaterialia, Vol. 6, No. 3, 2009, 359–367, ISSN 1214-9705.

Doležalová, H., Kajzar, V., Staš, L., and Souček, K. (2008) Observation of Subsidence Depression by Modern Geodetic Methods. Proceedings of SGEM 2008, Volume II. Albena, Bulgaria. p. 485-492. ISBN 954-918-181-2.

Mareš, S. et al. (1990) Introduction to Geophysics Praha: SNTL, 1990. 680 p. (in Czech).

Müller, K., Okál, M., and Hofrichterová, L. (1985) Principles of mining geophysics, Praha: SNTL/ ALFA. 224 p. (in Czech).

Chapter 9
Identification of Geotechnical Conditions in Areas of Former Shallow Mining Activity Using Geoelectrical Methods

Jolanta Pierwoła, Jarosław Badera, and Zbigniew Mirkowski

Abstract The inconvenient sinkholes in the area of the village of Kuźnica Masłońska were observed in last years. In the past, shallow exploitation of the Lower Jurassic brown coal was conducted in the vicinity of the village. The principal purpose of the study was to explain the origin of the deformations and to define the geotechnical conditions for building foundation. Geoelectrical investigations (resistivity and induced polarization imaging) and geotechnical identification were performed. Our identification showed the existence of a low resistive vertical zone near the surface, which is not connected with clayey sediments. According to geoelectrical measurements, this zone could be interpreted as a region of increased porosity and enhanced water infiltration. The region is connected with the former shaft. The geotechnical investigations pointed to unfavourable conditions caused by the existence of non-load-bearing beds and an intensive process of suffusion.

The conclusion about the partially post-mining character of the observed sinkholes intensified by adverse geological features and processes can be accepted.

9.1 Introduction

The effects of unknown exploitative voids have become a serious problem in many regions of former mining activity. The lack of areas for residential and industrial sites or highway engineering has created an interest in these previously unattractive areas. Unfortunately, in many cases the long-lasting mining activity, especially the oldest ones, is completely not catalogued or there are considerable shortcomings in the historical documentation.

One of the small former mining areas is situated near the city of Zawiercie (Poland). In this region, Lower Jurassic brown coal was extracted by several

J. Pierwoła (✉) · J. Badera · Z. Mirkowski
Department of Applied Geology, University of Silesia, Będzińska 60, 41 200 Sosnowiec, Poland
e-mail: jolanta.pierwola@us.edu.pl

A.F. Idziak and R. Dubiel (eds.), *Geophysics in Mining and Environmental Protection*, Geoplanet: Earth and Planetary Sciences 2, DOI 10.1007/978-3-642-19097-1_9, © Springer-Verlag Berlin Heidelberg 2011

underground mines from 1818 to 1959. The maximum intensity of mining works occurred in the 1920s, when 39 mines were excavated and about 240,000 tons of coal per year were extracted (Jaros 1975).

Some deformations, which are possibly connected with the above-mentioned mining operations, have recently been observed in the village of Kuźnica Masłońska, between Zawiercie and Łazy. In the past, four underground mines: Zygmunt (1919–1935), Stanisław (1926–1931), Kamilla (1910–1927, 1940–1944) and Ludwika (1891–1905, 1919–1924) operated in this area. In addition to the mines worked legally, there existed many small shafts where the local population extracted the coal illegally. A large number of these little "mines" were located close to the research area.

An explanation for the origin of the observed deformations, especially the role of the former mining works and the determination of geotechnical conditions of building foundations in the studied area, were the basic aims of the presented research. Electrical resistivity imaging and induced polarization methods were chosen for an initial identification of the rock-mass structure.

9.2 Geological Characteristics of the Research Area

The research area (Fig. 9.1) is situated on the border of the Silesia-Cracow Monocline, where Lower Jurassic and Triassic sediments appear on the surface. Near Zawiercie, the Upper Triassic (T_3) only occurs locally and it consists of red clay shale and mudstone with sandstone and gravel interlayers (Kotlicki 1967). The

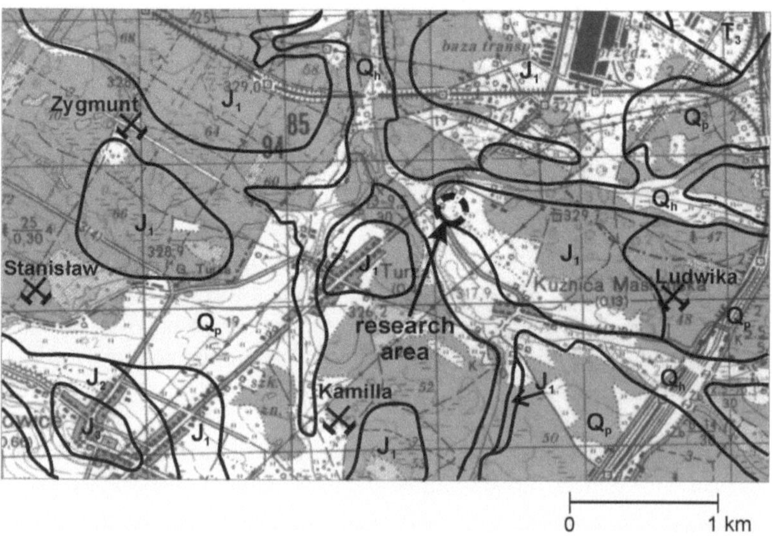

Fig. 9.1 Geological map of the research area with the location of former coal mines (according to Kotlicki 1966); notation explained in the text

formation of the Lower Jurassic (J_1) exists as isolated, erosionally-cut planes with variable thicknesses. These are different-grained sediments: gravels, sands, silt and clay, which were formed in an alluvial or lake environment (Kopik 1998). In the research area the above-mentioned formation consists of grey silt and clay with thin interlayers of fine-grained sand. These sediments, the so-called "Blanowicki strata", belong to the Pliensbach stage. A distinctive feature of this formation is the occurrence of sub-bituminous coal seams. The Lower Jurassic coal deposit is represented by the main coal seam 0.5–1.2 m in thickness and some thinner layers (0.05–0.3 m) situated above and below the main seam. The coal is black and bright, with low ash and high total sulfur contents and a high caloric value.

Above the Lower Jurassic, Middle (J_2) and Upper Jurassic (J_3) limestones can be found in some places.

On the majority of the area, the Triassic-Jurassic rocks are covered by Quaternary sediments. The Pleistocene (Q_p) occurs as a cover of perifluvioglacial sands of varying thicknesses and the Holocene (Q_h), connected only with a current stream valley, consists of alluvial clay and silt.

9.3   The Area of Investigation

The unforeseen appearance of sinkholes in the built-up area was the direct reason of undertaking the detailed research. The locations of sinkholes are presented in Fig. 9.2. The observed linear orientation of sinkholes suggests that they can be connected with the former mining activity. According to verbal information, a shallow, illegal exploitation of brown coal was conducted there. A thin coal seam with a thickness of about 0.8 m was extracted from a depth of about 5 m. The last sinkhole noted was about 1.5 m in diameter, 0.5 m deep and was limited by open fissures to a depth of minimum 2.5 m.

On account of the numerous buildings in the area, only a small surface was accessible for geophysical investigation. The research field was almost flat and changes in the levels did not exceed one meter. It dipped slightly to the south, toward the Czarna Przemsza River, which is situated at a distance of approximately 200 m. Two survey lines with NNW-SSE orientation and lengths of 80 and 65 m were traced across the line of the observed sinkholes.

9.4   Geophysical Investigations

Resistivity and induced polarization (IP) imaging using the LUND System (ABEM) were applied to identify the rock structure of the investigated area. Field conditions for the geoelectrical study were rather difficult – the upper layer of soil was weakly compacted and cracked, which has a bearing on the grounding of the electrodes and the electrical contact. Our measurements were performed after fortnightly rainfalls,

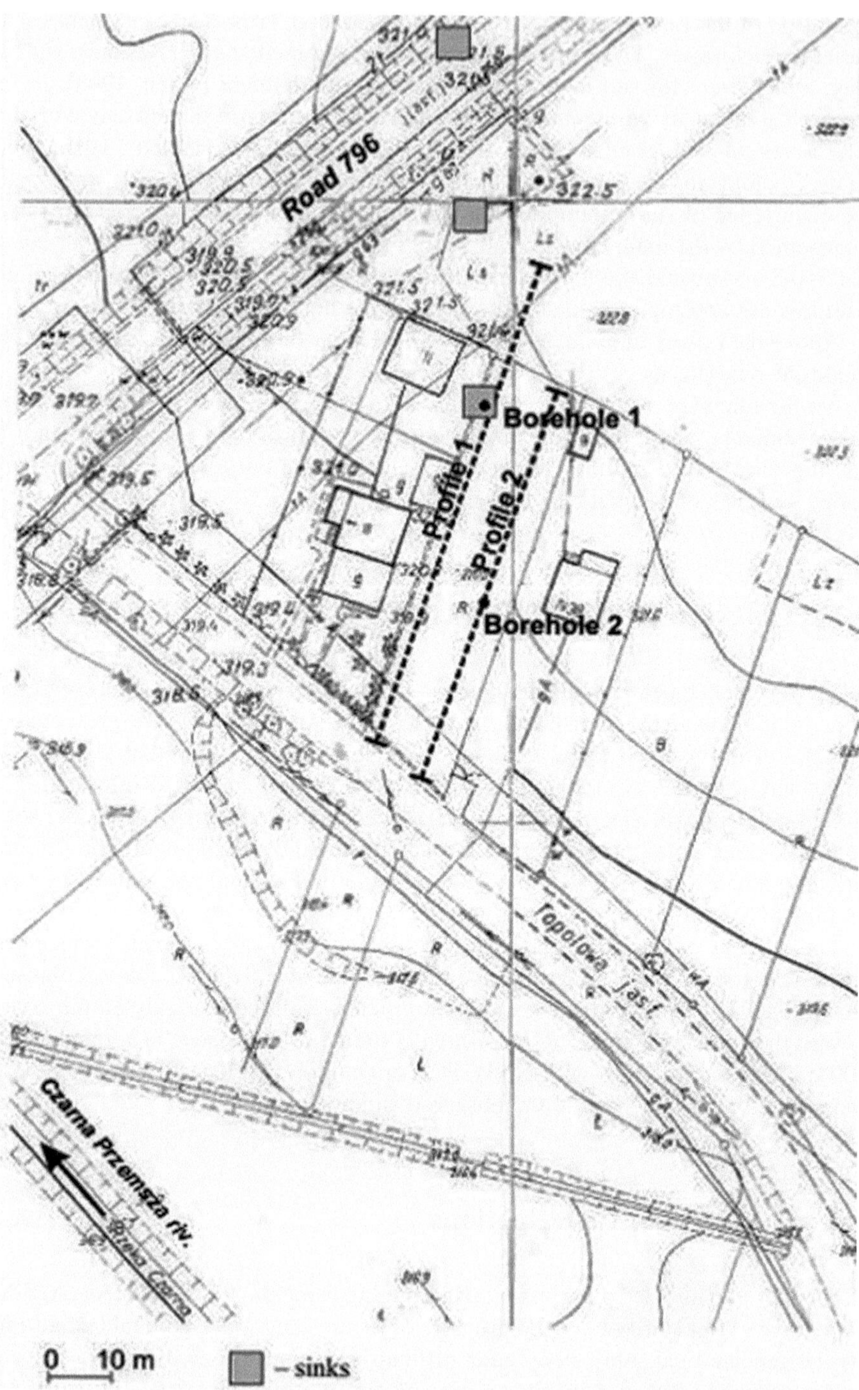

Fig. 9.2 Situational map of the examined area

which probably caused the cracks to be filled with water and this slightly improved the conditions for measurements.

The resistivity imaging method is often applied for the detection of underground voids, both natural (Nelson and Haigh 1990; Ezersky 2008; Roth et al. 2002; van Schoor 2002) and anthropogenic. In the last decade the IP method has become a useful tool that helps to distinguish between sediments of different lithological compositions (Slater and Lesmes 2002; Kemna et al. 2004) and hydraulic characteristics (Börner et al. 1996; Kemna et al. 2004).

The IP response depends on grain size, water content and water mineralization. It varies with the amount of water in the ground, increases if ground water has a low conductivity, and increases with decreasing porosity. The importance of rock texture for IP effects is characterised by a weak IP in compact clays and a strong IP measure in sediments with clay particles disseminated on the surface of larger mineral grains. The largest IP responses are obtained for clay contents of 3–10%.

In our study, the induced polarization was measured in the time domain where chargeability is the typically used measure of the IP effect. Chargeability represents a measure of polarization magnitude relative to conduction magnitude (Lesmes and Frye 2001; Slater and Lesmes 2002) and thus is approximately related to the bulk resistivity.

Resistivity and chargeability field measurements in Kuźnica Masłońska were performed with the dipole–dipole electrode configuration and an electrode distance of 5 m. The above-mentioned array was chosen because of its low electromagnetic coupling, which is important in an IP survey. The minimal emitted current intensity was set at 50 mA. The resistivity and chargeability data were inverted using RES2DINV software.

The depth of resistivity identification is about eight meters. That is slightly deeper than the expected depth of former mining works. The obtained resistivity values fluctuate within a range of 15 Ωm–3.8 kΩm (Fig. 9.3). Both sections exhibited a layered structure with a gradual decrease in resistivity with depth.

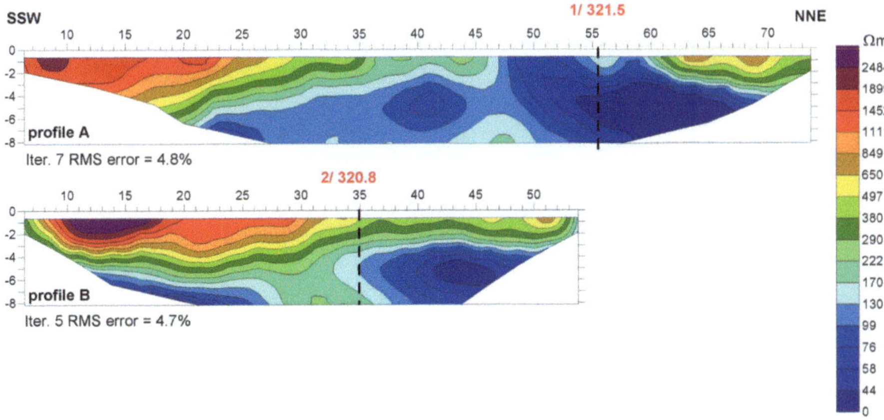

Fig. 9.3 The inverted resistivity cross-sections (*dashed lines* indicate boreholes)

A zone of very high resistivity values in both sections could be observed near the surface. We assume that this is connected with middle- and coarse-grained Quaternary or Lower Jurassic sediments. A shallow drilling, which was made on the 10th m of profile A, exhibited the existence of very dry sands to a depth of 0.5 m, which confirms our assumption.

Sediments with significantly lower resistivity values were found below sands at a depth of 4–6 m. The resistivities obtained are typical for layers with a clayey character or filled with water. In the central part of profile B, the resistivity is a bit higher than in the adjacent area, suggesting a change in that fraction toward coarse-grained.

The resistivity sections did not exhibit any zones of high resistivity, which might be connected with air-filled voids when considering the occurrence of unconsolidated rocks in the neighbourhood of the extracted seam, it is improbable that the voids existed for a longer period of time. Possibly they quickly got filled with unconsolidated surrounding sediments and water.

In profile A, between 47th and 59th m, it can be observed that the high resistivity zone is not continuous. In the centre of this interval (at 56th m), one of the sinkholes was situated, at present filled with sand. The presence of a low resistivity zone near the surface is thought to be due to the intensive infiltration of water connected with rainfalls.

The induced polarization measurement completed and disambiguated the resistivity results. The IP data obtained from greater depths were uninterpretable and thus the depth range for profile A is only 5 m. The IP sections presented in Fig. 9.4 showed that chargeability changes in the range of 0–144 ms.

The areas with significantly higher chargeability values are found at a depth of about 2.5–3 m in profile A, 4 m in the northern and 6 m in the southern part of profile B. The chargeability sections obtained were correlated with the resistivity sections. It can be observed that increased chargeability values correspond with decreased resistivity. Thus, it confirmed the clayey character of these sediments.

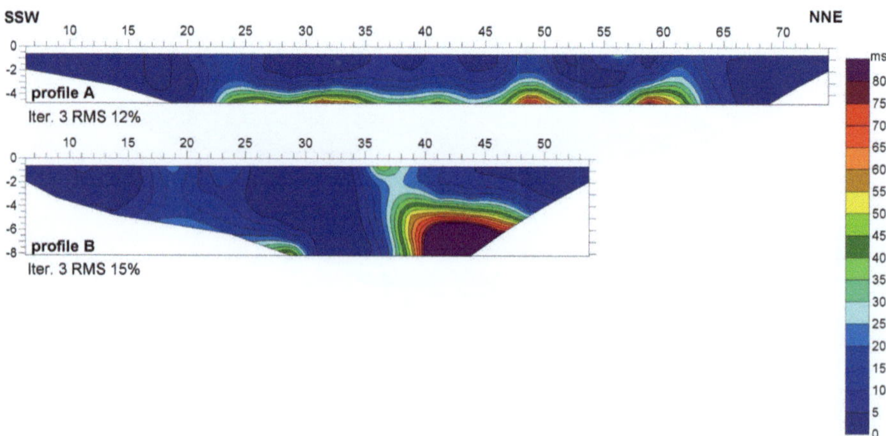

Fig. 9.4 Characterization of the examined area in terms of chargeability

The vertical, low-resistive zone, which was noticed near the surface on profile A (48–58 m), does not correlate with higher chargeability. On the contrary, one of the lowest chargeability values ($<$5 ms) was obtained in that interval. Because the chargeability increases with decreasing porosity, the low resistivity in combination with low chargeability can identify a part of rock-mass which is loosened and filled with water. In addition, the above-mentioned structure is situated in the vicinity of a filled-up sinkhole. This seems to confirm that sinkholes occur in the neighbourhood of former fore-shafts.

9.5 Geotechnical Research: Methods and Results

To determine the geotechnical conditions on the basis of geophysical measurements, two boreholes were made in the investigated area: the first one – directly within the sinkhole, the second – about 30 m to the south from the first one (Fig. 9.2). They reached the depths of 10 and 12 m, respectively. The macroscopic field description of the bored layers included: kind and colour of ground, state of ground, humidity and number of rollings. Furthermore, bored and stabilized ground water levels were registered.

Selected depth intervals were sampled for laboratory research. Laboratory measurements of grounds included a granulometric analysis, fixing of volume density (ρ) and degree of density (I_D) or degree of plasticity (I_L). Other parameters – apparent angle of internal friction (φ_u), apparent coherence (C_u), the modulus of total ground deformation (E_o) and the edometric modulus of original compressibility (M_o) – were established according to Polish standards (Table 9.1).

A cross-section containing lithological structure and hydrogeological conditions was constructed from the boreholes information (Fig. 9.5). Three main complexes of grounds have been sectioned – these are: recently created ground (I), Pleistocene alluvial-glacial sediments (II) and Lower Jurassic continental sediments (III).

Next, individual geological-engineering beds (a–e), which are lithologically and geotechnically diversified, have been set apart within these complexes (Table 9.1; Fig. 9.5). The Blanowice coal seams (0.2–1.5 m in thickness) are strongly weathered, fractured and soaked with water, so they form a non-load-bearing bed. Underlying sands have a character of floating earths. Load-bearing of overlying humus silts is also rather poor. The weathered brown coals and the Lower Jurassic sands form 1–3 aquifers. They lie at a depth of approximately 6 m. Groundwater has a confined surface and flows toward the Czarna Przemsza River, whose channel is about 200 m SSW of the area examined.

Generally, the geotechnical conditions turned out to be complicated – adverse lithology and high water-bearing capacity causes low load capacity of the ground and the development of the suffusion effect. This phenomenon consists in the mechanical rinsing of mineral grains off poorly compacted sediments, as a result of the infiltration of meteoric water in the rock basement and the migration of shallow groundwater (intensive because of the closeness of the channel of the

Table 9.1 Geotechnical parameters of grounds examined

Geotechnical bed	Type of ground	Volume density ρ [t/m^3]	Degree of density I_D	Degree of plasticity I_L	Apparent angle of internal friction $\varphi_u^{(n)}$ [º]	Apparent coherence $C_u^{(n)}$ [kPa]	Modulus of total ground deformation $E_o^{(n)}$ [MPa]	Edometric modulus of original compressibility $M_o^{(n)}$ [MPa]
I	N[Pd + Pπ]				Non-load-bearing bed			
II a	Pd, Pd//π	1.75	0.45		30.2		42.1	56.4
II b	Gπ	2.05		0.11	16.2	21.5	25.4	36.3
III a	H(π//Gπ), Hπ	1.85		0.42	9.5	9.4	11.6	16.9
III b	Pπ/π + K	1.85	0.35		29.7		34.7	46.6
III c	Ps + Z + K	1.85	0.45		32.7		73.2	86.7
III d	I, Iπ, Iπ//Pd, Iπ//C-b	2.08		0.05	12.3	57.1	19.6	34.6
III e	C-b				Non-load-bearing bed			

Gb soil; *N* made ground; *K* rock chips; *Z* gravel; *Ps* medium-grained sand; *Pd* fine-grained sand; *Pπ* silty sand; *π* silt; *I* clay; *Iπ* silty clay; *Hπ* humus silt; *Gπ* silty loam; *C-b* brown coal; + additives; // intralayers; / intermediate grounds; [n] data based on Polish standards

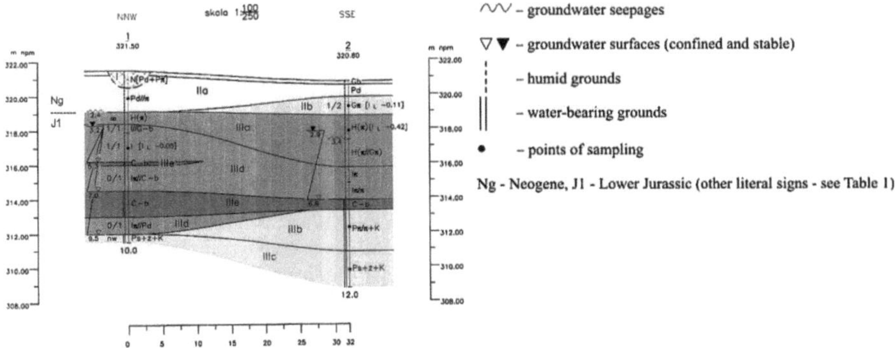

Fig. 9.5   Geological-engineering cross-section in the examined area

Czarna Przemsza River). Additionally, ground parameters may deteriorate under the influence of static (buildings) and dynamic (traffic) loads, faulty drainage of buildings and road facilities, and the effects of former mining.

9.6   Conclusions

1. The combination of resistivity and induced polarization imaging appeared to be a useful tool for identification of areas of former mining activity.
2. Our investigations confirmed the partial post-mining character of sinkholes occurring in the examined area.
3. The geoelectrical measurements did not show any high resistivity zones in the examined area, which might be connected with post-exploitation voids (filled with air). No underground voids appeared during drilling.
4. Plasticity and the high water-bearing capacity of layers (enclosing coal seams) could be the reason for the quick clenching of excavations that existed in the past.
5. Most probably, suffusion is a direct cause of the occurrence of sinkholes.
6. The rock material in the vicinity of supposed excavations is probably mixed; the coal itself is weathered and fractured. Owing to this fact, water migration becomes more intense and suffusion is induced as a result. This especially concerns former shafts which are places of increased infiltration of meteoric water.

References

Börner F., Schopper J., and Weller A. (1996) Evaluation of transport and storage properties in the soil and groundwater zone from induced polarization measurements; Geophys. Prospect., **44**, 583–601.

Ezersky M. (2008) Geoelectric structure of the Ein Gedi sinkhole occurrence site at the Dead Sea shore in Israel Journal of Applied Geophysics, **64**, 56–69.

Jaros J. (1975) Zarys dziejów górnictwa węglowego [Outline of the Mining History]. PWN Warszawa-Kraków (in Polish).

Kemna A., Binley A., and Slater L. (2004) Crosshole IP imaging for engineering and environmental applications; Geophysics, **69**, 97–107.

Kopik J. (1998) Jura dolna i środkowa północno-wschodniego obrzeżenia GZW [Upper and Middler Jurassic of the NE margin of the Upper Silesian Coal Basin]. Biuletyn PIG, **378** (in Polish).

Kotlicki S. (1966) Szczegółowa mapa geologiczna Polski. Arkusz Zawiercie [Detailed geological map of Poland. Zawiercie] 1:50000, PIG Warszawa (in Polish).

Kotlicki S. (1967) Objaśnienia do szczegółowej mapy geologicznej Polski. Arkusz Zawiercie. [Explanation to detailed geological map of Poland], PIG Warszawa (in Polish).

Lesmes D., and Frye K. (2001) Influence of pore fluid chemistry on the complex conductivity and induced polarization responses of Berea sandstones; J. Geophys. Res., **106**, 4079–4090.

Nelson R.G., and Haigh J.H. (1990) Geophysical investigations of sinkholes in lateritic terrains – Geotechnical and environmental geophysics., ed. S. Ward, v.3, 133–153.

Roth M., Mackey J.R., Mackey C., and Nyquist J.E. (2002) - A case study of the reliability of multielectrode earth resistivity testing for geotechnical investigations in karst terrains, Engineering Geology, **65**, 225–232.

Slater L., and Lesmes D. (2002) IP interpretation in environmental investigations; Geophysics, **67**, 68–77.

van Schoor M. (2002) Detection of sinkholes using 2D electrical resistivity imaging, Journal of Applied Geophysics, **50**, 393–399

Chapter 10
The Use of Seismic Methods to Determine Potential Directions of Underground Water Flow in a Fractured Rock Mass

Adam Idziak and Iwona Stan-Kleczek

Abstract In fractured rock mass, the orientation of cracks causes anisotropy of different physical properties, including the permeability coefficient. In rocks where the permeability of rock matrix may be negligible, flow occurs mainly through the connected fractures, or fracture networks. The rock mass which contains fractures of this kind can be treated as a homogeneous anisotropic porous medium to formulate the permeability tensor. This can be possible owing to introduction of crack tensor which depends only on the aperture, size and orientation of the related cracks. The crack tensor is coaxial with the permeability tensor. The relation between the permeability tensor and the crack tensor enables us to utilize the measurement of anisotropy of seismic wave velocity to establish potential directions of underground water flow.

10.1 Introduction

In the present paper, relationship between the crack tensor, seismic wave velocity tensor and permeability tensor established for fractured rocks is described. In the earlier papers (Idziak 1995; Stan-Kleczek 2008), the relation between the tensors was studied for fractured carbonate rocks. In this paper, we are trying to check this relation for another kind of rocks.

Fractures are one of the main features determining the rock permeability, which is, the capability of rock mass to be infiltrated by fluid, gas and/or their mixture under a pressure gradient. It depends on a hydraulic network occurring in the rock and a character of fluid or gas motion. In carbonate rocks, the hydraulic network consists of three types of discontinuities:

A. Idziak • I. Stan-Kleczek (✉)
University of Silesia, 60 Bedzinska Street, Sosnowiec, Poland
e-mail: istan@us.edu.pl

A.F. Idziak and R. Dubiel (eds.), *Geophysics in Mining and Environmental Protection*,
Geoplanet: Earth and Planetary Sciences 2, DOI 10.1007/978-3-642-19097-1_10,
© Springer-Verlag Berlin Heidelberg 2011

- Spherical pores which are uniformly spaced in a rock matrix.
- Fractures which form regular discontinuity network in a rock matrix.
- Karst forms of irregular and different shape, frequently of big size.

The motion of a fluid in pores and fractures is subject to linear Darcy's law, whereas in karst the channel flow is nonlinear by dint of big velocity of fluid.

10.2 Theoretical Model

10.2.1 Permeability Tensor

The permeability factor depends on flow direction so it must be described in a tensorial form. The rank of permeability tensor has to match the complexity of flow anisotropy. In the simplest case, the permeability tensor can be of rank two, in more complex cases it can be of rank four or bigger.

The knowledge of permeability tensor enables us to determine a directional permeability factor in the direction described by the unit vector $\vec{n}$:

$$K(\vec{n}) = \hat{K}_{ij...z} \cdot n_i n_j ... n_z, \tag{10.1}$$

where $\hat{K}_{ij...z}$ are components of permeability tensor of rock mass, and n_k are directional cosines of vector $\vec{n}$. Above some critical crack density in a rock mass, one can treat the fractured medium as an equivalent anisotropic porous medium (Idziak 1995) and describe fluid flow by Darcy's law.

The flow velocity in the direction described by unit vector $\vec{n}$ can be calculated from:

$$v(\vec{n}) = -\frac{g}{v} K(\vec{n}) \vec{J} \vec{n}, \tag{10.2}$$

where g is gravitational acceleration, v the kinematic viscosity of the fluid, $\vec{J}$ the vector of hydraulic gradient. Scalar product $\vec{J} \cdot \vec{n}$ is a projection of hydraulic gradient vector $\vec{J}$ on direction $\vec{n}$.

10.2.2 Crack Tensor

To describe the crack anisotropy, a tensor quantity called *crack tensor* can be used (Idziak and Stan-Kłeczek 2006). The crack tensor of rank k can be defined as:

$$\hat{F}_{ij...l} = \left\langle \varepsilon A_{ij...l} \right\rangle, \tag{10.3}$$

where $A_{ij...l}$ is the cartesian product of the unit vector $\vec{n}$ normal to the principal plane of crack, and ε is a crack density parameter.

To estimate the crack tensor components one has to know a statistical distribution of crack geometry and orientation, represented by density function $E(a, c, \vec{n})$ where a is diameter of crack and c is its aperture. The function has to be invariant to the sense of vector $\vec{n}$:

$$E(a, c, \vec{n}) = E(a, c, -\vec{n})\,. \tag{10.4}$$

Estimators of the crack tensor components for penny-shape cracks can be calculated from the relation:

$$\hat{F}_{i,j...l} = \frac{\pi}{4} \int\limits_{0}^{a_{max}} \int\limits_{0}^{c_{max}} \int\limits_{\Omega} a^2 c n_i n_j ... n_l E(a, c, \vec{n})\, da\, dc\, d\Omega\,, \tag{10.5}$$

where Ω is the solid angle corresponding to the entire surface of the unit sphere.

10.2.3   Calculation of Permeability Tensor Based on Crack Tensor

To establish relationship between crack and permeability tensors it is necessary to take into account critical crack density below which connections between fractures are broken up and the fracture permeability is equal to zero (Oda 1985). The relationship between tensors can be described by formula:

$$\hat{K}_{ij} = G \cdot (\tilde{\varepsilon} \cdot \delta_{ij} - \hat{F}_{ij}) \tag{10.6}$$

where $\tilde{\varepsilon}$ is the sum of crack tensor eigenvalues (a trace of tensor matrix), and G is a scalar factor depending only on crack geometry and critical density of cracks. Both tensors, $\hat{K}_{ij}$ and $\hat{F}_{ij}$, are coaxial; the longer axis of permeability tensor corresponds to the shorter axis of crack tensor and the shorter axis of permeability tensor corresponds to the longer axis of crack tensor.

To check this relation, four models for two crack systems cutting by different angles were created (Figs. 10.1–10.4).

10.2.4   Velocity Tensor

Tensor calculus can be applied to describe directional distribution of seismic wave velocity. Velocity tensor determined on the basis of seismic velocity measurement maps seismic anisotropy and allows to calculate velocity in any direction. Velocity

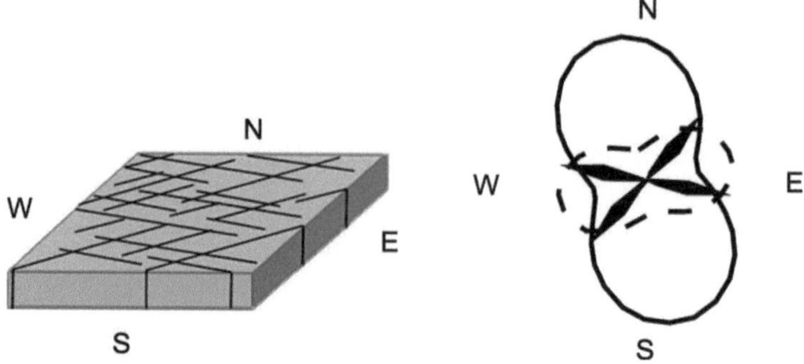

Fig. 10.1 The relation between geometrical representation of crack tensor (*solid line*) and permeability tensor (*dashed line*) for two systems with the same crack densities cutting by 60°

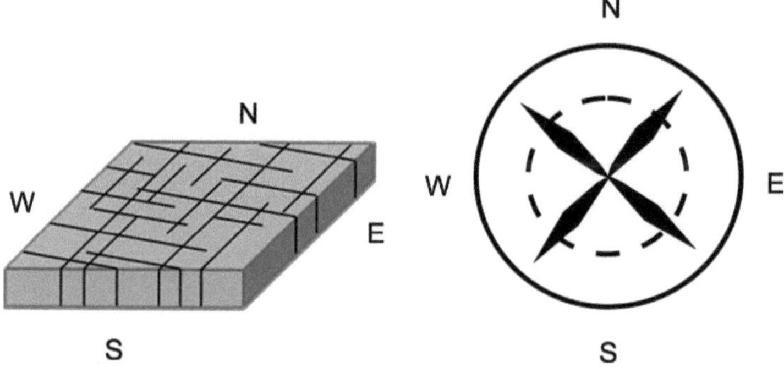

Fig. 10.2 Model showing the relation between crack tensor (*solid line*) and permeability tensor (*dashed line*) for two systems with the same crack densities cutting by 90°

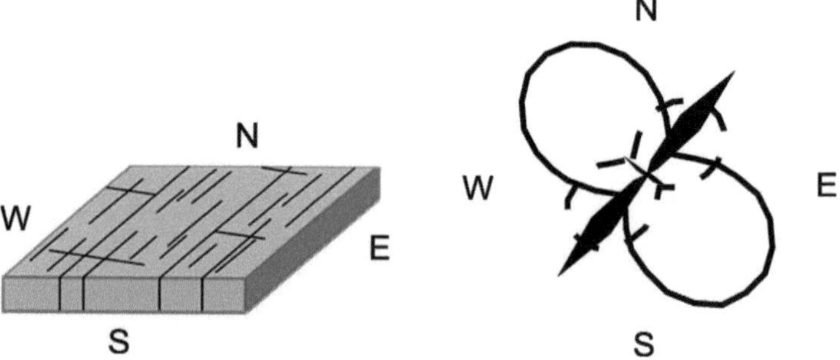

Fig. 10.3 Model showing the relation between crack tensor (*solid line*) and permeability tensor (*dashed line*) for two systems with different crack densities cutting by 90°

tensor can be calculated from velocity measurements in N independent directions described by unit vectors $\mathbf{m}^{(p)}$ (Stan and Idziak 2005). The tensor components $V_{l_1 l_2 \ldots l_k}$ are calculated from the set of equations:

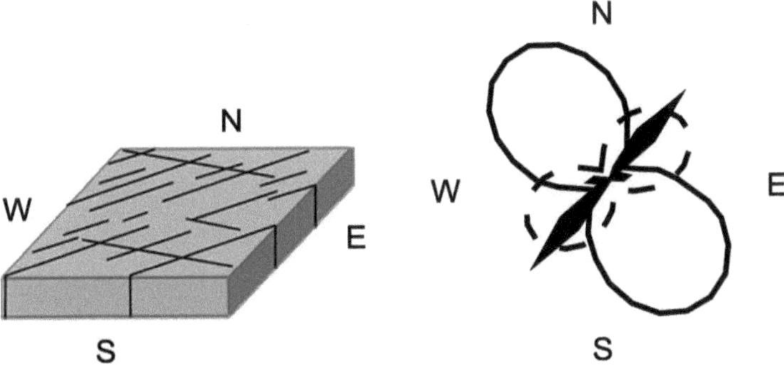

Fig. 10.4 Model showing the relation between crack tensor (*solid line*) and permeability tensor (*dashed line*) for two systems with different crack densities cutting by 60°

$$\left(\frac{v_p}{v_0}\right)^2 = V_{l_1 l_2 \ldots l_k} \cdot m_{l_1}^{(p)} m_{l_2}^{(p)} \ldots m_{l_k}^{(p)} \quad \mathrm{p} = 1, 2 \ldots \mathrm{N} \tag{10.7}$$

where $m_l^{(p)}$ are directional cosines of the vector $\mathbf{m}^{(p)}$. The wave velocity v_0 is used as a normalizing factor. It can be a seismic velocity in unfractured rock mass or a maximum measured velocity.

10.2.5 *Estimation of Permeability Tensor from Seismic Measurements*

Properly designed seismic measurements deliver information about wave velocity for different directions of wave propagation. The obtained distribution of P-wave velocity allows to make qualitative and quantitative analysis of crack anisotropy. The relationships between crack and velocity anisotropy in different rock mass were shown in earlier publications (Stan and Idziak 2005; Idziak and Stan-Kłeczek 2006; Stan-Kleczek 2008).

In the case of crack perpendicular to the rock bedding, both crack and velocity tensors are two-dimensional. Rotating eigenvectors of velocity tensor by 90° one can obtain tensor $\hat{U}$ whose components are:

$$U_{11} = V_{22} \quad U_{22} = V_{11} \quad U_{12} = U_{21} = -V_{12} \tag{10.8}$$

The tensor represents directional properties of the crack system. To calculate components of crack tensor, the relationships between mean values and anisotropy coefficients of crack and velocity tensors have to be known (Idziak 1995). Eigenvalues of the crack tensor can be then estimated as:

$$f1 = \tfrac{1}{2}\left(1 - K_F\right)\ \Psi(A_V) \quad f2 = \tfrac{1}{2}\left(1 + K_F\right)\ \Psi(A_V) \tag{10.9}$$

where K_F is the crack anisotropy coefficient, $\Psi(A_V)$ the function describing the dependence of mean value of crack tensor on mean value of velocity tensor $\mathbf{A_V}$.

The obtained components of the crack tensor can be used to estimate the permeability tensor components accordingly to (10.6).

10.3 The Field Study and Results

The direct mesostructure observations and seismic measurements were made in quarries located in southern Poland. The present study concerns sedimentary rocks like sandstone (deposit 1) and igneous rocks like diabase (deposit 2). The strike azimuth and dip angle were measured using the geological compass. The rose diagrams of crack distribution were drawn on the basis of crack orientation data. The results of crack systems orientation and their geometrical parameters measurements were utilized to calculate the crack tensor. The seismic equipment Terraloc MK6 was used for recording of seismic waves in surface layers of the rock mass. The measurements were made along precisely oriented radial seismic profiles. The azimuth interval between profiles was 10°. The seismic data were digitally recorded with up to 12 geophones at a 3-m spacing, so the seismic spreads were 33-m long. Seismic waves were generated by an 8-kg hammer which was hit against a metal plate. The accuracy of seismic velocity measurement depended on the precision of apparatus and local wave velocity changes along the profile. The first breaks of P waves were read from recorded seismograms. Wave velocities were calculated from a slope of linear refraction travel time-offset relation obtained by least-square fitting to experimental data. The surface orientation diagrams were drawn on the basis of crack orientation data. Permeability tensors were calculated from (10.6) where G was equal to 0.6, assuming that above 60% of cracks participate in a fluid flow. The axes of permeability tensors were rotated by 90° in relation to the axes of crack tensors. The results are shown in Figs. 10.5a and 10.6a. The results of seismic anisotropy measurements are presented in Figs. 10.5b and 10.6b.

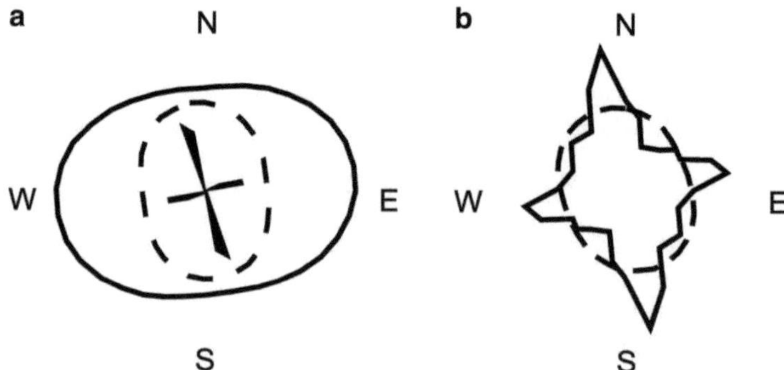

Fig. 10.5 Deposit 1: (**a**) spatial distributions of cracks based on rose diagrams of crack orientation, crack tensor of rank two (*solid line*) and permeability tensor of rank two (*dashed line*), (**b**) relationship between the azimuth of seismic profile and P-wave squared velocity (*solid line*) as well as velocity distribution based on the tensor of rank two (*dashed line*)

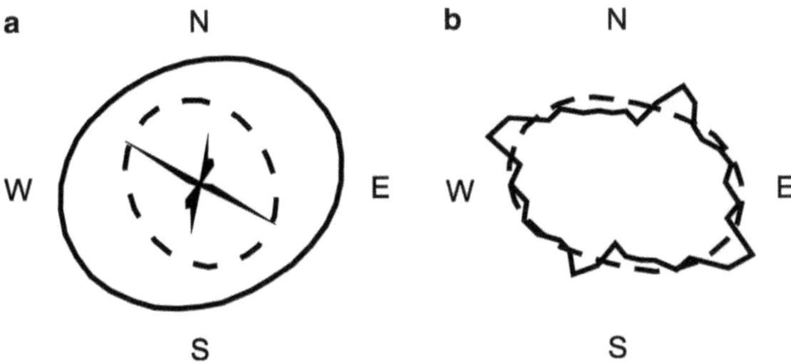

Fig. 10.6 Deposit 2: (**a**) spatial distributions of cracks based on rose diagrams of crack orientation, crack tensor of rank two (*solid line*) and permeability tensor of rank two (*dashed line*), (**b**) relationship between the azimuth of seismic profile and P-wave square velocity (*solid line*), velocity distribution based on the tensor of rank two (*dashed line*)

10.4   Conclusions

The presented study confirmed that according to the assumed model of flow in fractured rock, the velocity tensor and the permeability tensor are coaxial when the two existing crack systems are almost perpendicular (Fig. 10.5). The longer main axes of velocity tensor and permeability tensor are parallel. This situation was observed at the site 1. At the site 2, besides two perpendicular crack systems, a third system close to one of them exists. This additional system with a small crack density caused that the axes of permeability tensor calculated on the basis of crack tensor differed slightly from the axes of velocity tensor but the azimuth distribution of wave velocity shows that the third system has no meaning for velocity anisotropy. It is possible that because of its small density this system did not affect the fluid conductivity.

The presented theory concerns only flattened and partially fluid-filled cracks. The relation between permeability tensor and velocity tensor may enable the use of measurements of seismic wave velocity anisotropy in fractured rocks for studying their hydraulic properties. This study showed that measurements of seismic anisotropy can be a useful tool in hydrogeological research of fractured aquifers.

References

Idziak A.F. (1995). *Wykorzystanie pomiarów anizotropii sejsmicznej i anizotropii oporu elektrycznego do badania dróg przepływu w masywach szczelinowatych*, [The use of seismic anisotropy and electric resistivity measurements in the study of flow paths in fractured rock masses],Współczesne problem hydrogeologii t. VII, Kraków-Krynica (in Polish).

Idziak A.F. and Stan-Kłeczek I. (2006) *Geomechanical properties of fractured carbonate rock mass determined by geophysical methods*, in van Cotthem A. (ed): Multiphysics Coupling and Long Term Behavior in Rock Mechanics, Taylor & Francis/Balkema, London, 551–556.

Oda M. (1985) *Permeability tensor for discontinuous rock masses*, Geotechnique, 35, no 444, 483–495.

Stan I. and Idziak A. (2005) *Anisotropy of seismic waves velocity due to the fracturing in chosen rock mass*; in P. Konečny (ed.), Impact of Human Activity on the Geological Environment, A.A. Balcema., London, 579–586.

Stan-Kleczek I. (2008) *The role of seismic methods in investigation of rock mass*, Acta Geophysica, vol. 56, no. 4, 1065–1073.

Chapter 11
Assessment of Fracture Properties from P-Wave Velocity Distribution

Jan Vilhelm, Vladimír Rudajev, and Roman Živor

Abstract Seismic measurements carried out on peridotite rock outcrop showed that seismic wave velocity depends on the direction of propagation. It was found out that the velocity is influenced on system of fractures. The fracture sets can cause velocity dispersion. The theory of displacement discontinuity was applied for an explanation of fractures influence on the seismic wave propagation. According to this theory, the fracture stiffness can be assessed on the basis of velocity directional dependence. The measurements made proved experimentally the validity of this approach, specifically for the frequency interval from 500 Hz to 700 kHz.

11.1 Introduction

Seismic field measurements frequently display directional dependence of propagation velocity (Přikryl et al., 2004). This can be caused by rock anisotropy, on the one hand, and by the fracturing of rocks, on the other (Musgrave, 1970).

The elastic anisotropy of rocks can be studied conveniently in the laboratory, observing, e.g., the effect of the mineral orientations or crystallographic structure on the propagation of ultrasonic waves. Since the wavelengths of ultrasonic waves are larger than the micro fractures contained in the rock sample, the observed laboratory velocity values can be assumed, in the first approximation, to be the velocities in a continuous unperturbed medium. Usual experiments work only with longitudinal waves (P-waves).

J. Vilhelm (✉)
Charles University in Prague, Albertov 6, 128 43 Praha 2, Czech Republic
and
Institute of Geology AS CR, v.v.i., Puškinovo náměstí 9, 160 00 Praha 6, Czech Republic
e-mail: vilhelm@natur.cuni.cz

V. Rudajev • R. Živor
Institute of Geology AS CR, v.v.i., Puškinovo náměstí 9, 160 00 Praha 6, Czech Republic

A.F. Idziak and R. Dubiel (eds.), *Geophysics in Mining and Environmental Protection*, Geoplanet: Earth and Planetary Sciences 2, DOI 10.1007/978-3-642-19097-1_11,

The systems of fractures in a rock massif cause not only directional dependence of P-wave velocities, but also dispersion, i.e., the dependence of seismic wave velocity on frequency. Systems of closed and open fractures have different effects. The research into the parameters of fracture systems is important not only in assessing the mechanical properties of the rock medium, but also for the assessment of their permeability and fluid saturation.

11.2   Method of Assessing Fracture Parameters

The velocity of propagation of seismic waves in a medium weakened by fractures was analysed using the theory of displacement discontinuity (Pyrak-Nolte et al., 1990). This approach is founded on deriving the effect of the presence of one plane fracture on a propagating elastic wave. In describing the behaviour of the continuum with the fracture, it is evident that the fracture will display a larger relative deformation than the ambient unperturbed medium. Due to the acting stress, the fracture may close. In a particular interval of fracture deformations, the specific fracture stiffness can be defined. The specific fracture stiffness is, therefore, defined as the ratio of the stress increment across the fracture to the corresponding strain change. The fracture stiffness is thus a physical quantity which describes the fracture's elastic behaviour. That is why it can be used in deriving the properties of an elastic wave propagating through a medium with a fracture. The reciprocal value of the fracture stiffness is the fracture compliance.

Solving the equation of propagation of an elastic wave through a medium with a fracture is founded on introducing boundary conditions at both plane edges of the fracture. It is assumed that the fracture has the nature of a non-welded contact, which causes the displacement at the fracture to be discontinuous, although the stress field is continuous. The discontinuity of the displacement perpendicular to the fracture occurs at both boundary planes of the fracture, and is proportional to the stress along the normal to the fracture. The result is that the transmission coefficient T for the P-wave under normal incidence, with equal properties on either side of the fracture, becomes frequency dependent:

$$T(\omega) = \frac{2\frac{\kappa}{Z}}{-i\omega + 2\frac{\kappa}{Z}},$$

(11.1)

where κ is the stiffness (in the z direction, i.e., perpendicularly to the fracture); Z is seismic impedance; ω is frequency.

For limiting cases of fracture stiffness κ, transmission coefficient T takes the following limiting values:

- $\lim\limits_{\kappa \to \infty} T = 1$: the solution reverts to the welded contact, and the fracture has no effect.

- $\lim\limits_{\kappa \to 0} T = 0$: the solution reverts to the case of a free surface, and all energy is reflected.

The transmission coefficient in the medium with the fracture is a complex quantity. As the wave propagates through the fracture, phase delay occurs. By simple modification of the expression for the reflection coefficient, its real and imaginary parts can be determined:

$$T(\omega) = \frac{2\frac{\kappa}{Z}}{-i\omega + 2\frac{\kappa}{Z}} = \frac{4\left(\frac{\kappa}{Z}\right)^2}{\omega^2 + 4\left(\frac{\kappa}{Z}\right)^2} + i\frac{2\omega\frac{\kappa}{Z}}{\omega^2 + 4\left(\frac{\kappa}{Z}\right)^2}. \tag{11.2}$$

Phase delay Θ is then:

$$\Theta(\omega) = arctg\left(\frac{\omega}{2\frac{\kappa}{Z}}\right). \tag{11.3}$$

The transmitted wave becomes dispersive and its group velocity can be determined. The group time delay t_g of single fracture is then:

$$t_g = \frac{d\Theta}{d\omega} = \frac{2\frac{\kappa}{Z}}{\left[4\left(\frac{\kappa}{Z}\right)^2 + \omega^2\right]}. \tag{11.4}$$

This relation holds for normal propagation of waves across one fracture. However, it can be generalised for any angle of incidence of the waves and can also be expanded for multiple parallel fractures. The fracture density is expressed in terms of the number of fractures N per unit length L. The selected approach can be simplified by neglecting multiple reflections from separate fractures and by considering no interaction between the separate fractures. This simplification is real if the distance between the separate parallel fractures is larger than the distance between the asperities. The effect of a number of parallel fractures may then be expressed as the sum of the delays at the individual fractures. The effective group velocity U for normal incidence of waves on a medium perturbed by parallel fractures may be calculated using the relation:

$$U = \frac{V\left\{1 + \left[\frac{\omega}{2\frac{\kappa}{Z}}\right]^2\right\}}{1 + \left[\frac{\omega}{2\frac{\kappa}{Z}}\right]^2 + \frac{NVZ}{2L\kappa}}, \tag{11.5}$$

where V is the velocity in non-disturbed and non-dispersive medium (phase velocity), L/N is the average distance between neighbouring fractures, κ is the fracture stiffness and Z is the seismic impedance ($Z = \rho.U$, where ρ is the rock density). An example of group velocity dependence on frequency is given in Fig. 11.1.

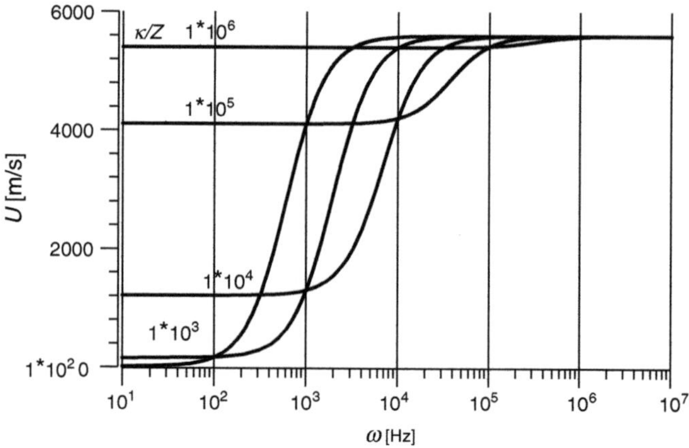

Fig. 11.1 Group velocity as a function of frequency ω for different ratios κ/Z. Parameters according to Pyrak-Nolte et al. (1990): $V = 5{,}600$ [m/s], $\rho = 2{,}600$ [kg/m^3], $L/N = 0.077$ [m/fracture], $k/Z - $ [1/s]

Since the direct measurement of fractures stiffness κ of fractures in real rocks is very difficult, its value is estimated from seismic measurements (Worthington, 2008).

This paper describes the determination of fracture stiffness using a combination of laboratory and field measurements of seismic wave velocities. The measurements were made in many directions and made use of an extraordinarily broad interval of elastic wave frequencies, used to calculate the seismic wave velocities.

11.3   Laboratory and Field Measurements

The objects of the research were systems of fractures observed in the ultrabasic rocks of the Ivrea Zone (north-western Italy). These rocks represent material, the composition and origin of which correspond to depths of the lower crust and upper mantle, and good knowledge of their seismic properties may help in producing more accurate seismic models of this depth zone.

The laboratory velocity measurements were made by the pulse-transmission technique on spherical rock specimens (5 cm in diameter) according to method published by Pros and Podroušková (1974) and Pros (1977). The method makes it possible to measure in many directions. The frequency of the sounding pulse was 700 kHz. The sampling frequency of the recorded signals was 100 MHz.

Only velocities corresponding to the orientation of the outcrop surface plane in the field determined under laboratory conditions at atmospheric pressure (0.1 MPa) were used for comparison with field data and namely for fracture stiffness determination.

The laboratory velocities displayed practically no anisotropy in the plane of the outcrop; the difference between the lowest and the highest velocities was 110 m/s, which, given the average velocity determined in this plane of approx. 7,730 m/s, is less than 1.5% of this velocity.

The seismic field measurements were carried out with a 24-channel Geode engineering seismograph using a hammer as the seismic source. The frequency of the recorded P-waves was between 0.5 and 1.8 kHz.

Three systems of sub-vertical fractures were observed macroscopically at the surface of the outcrop. The first system is oriented from E to W. The fractures of this system are opened at the surface and the distance between neighbouring parallel fractures is about 20 cm. Only fractures of this system (E-W) could be observed continuously up to a distance of several metres. The second system of fractures, running N to S, consisted of tight fractures with a density of as much as 1 fracture per 10 cm in places. The third system of fractures was a system running from SW to NE. These fractures were tight and the distance between neighbouring cracks mostly exceeded 50 cm.

The anisotropy determined by field measurements reflects the properties of the observed sub-vertical sets of parallel fractures. The minimum velocity values were observed perpendicular to the system of open fractures, running from E to W. On the contrary, the highest velocities were identified in the direction of this system (i.e., in the direction E-W). The observed set of fractures oriented in the direction N-S could influence mostly the velocity in the direction E-W. The orientation of the highest velocity observed, however, is in agreement with the observed tightness of the N-S fracture set, even with its high density (0.1 m/fracture). Sparse (0.5 m/fracture), tight fractures, running in the NE-SW direction only slightly slowed down the velocity of propagation of seismic waves in the perpendicular direction (NW-SE).

11.4   Interpretation of Laboratory and Field Data

The influence of sets of parallel fractures on the transmission of seismic waves was evaluated by means of the above displacement discontinuity model.

The velocity determined by laboratory measurement was regarded as velocity value V in the undisturbed medium ($V = 7,800$ m/s) and it was substituted into the (11.5). This enabled to apply displacement discontinuity method separately to individual systems of parallel fractures. The values of the velocities and fracture densities in the separate fracture systems, determined in the field, were used to calculate fracture stiffness κ. The density of peridotite–lherzolite was $\rho = 3,500$ kg/m^3.

The obtained values of the fracture stiffness are demonstrated in Fig. 11.2 (curves A). The graph of the frequency dependence of the group velocity U provides an explanation of the difference in velocities in the field (frequencies in the interval of 500–1,000 Hz, in the extreme up to 1,800 Hz) and in the laboratory

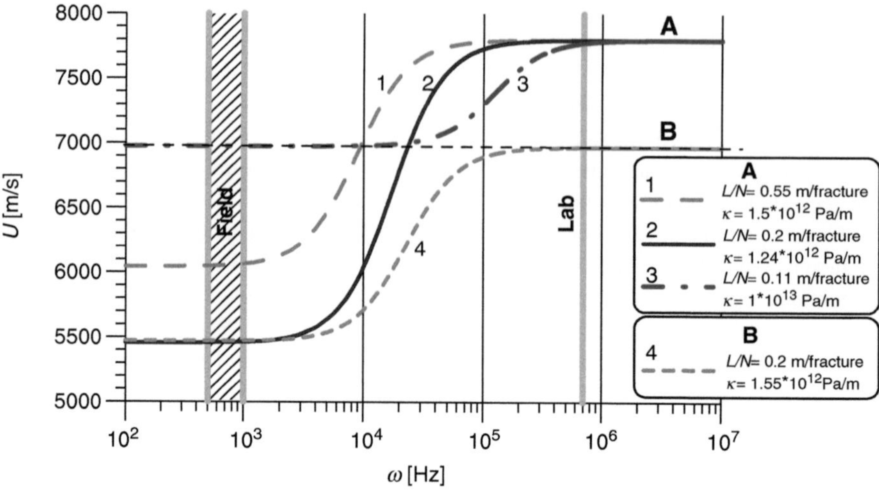

Fig. 11.2 Velocity as a function of frequency, normal incidence (according to Pyrak-Nolte et al., 1990). *A* fracture stiffness calculated independently for all the three fracture sets, based on laboratory determined velocity in undisturbed medium. *B* fracture stiffness of E-W fracture set, determined from field data only. In this case, the other fracture sets (N-S, SW-NE) are expected to be too stiff and all the anisotropy is due to E-W fractures

(700 kHz). The system of open E-W fractures corresponds to the lowest value of the fracture stiffness ($\kappa = 1.24 \times 10^{12}$ Pa/m). On the contrary, the highest value of the fracture stiffness ($\kappa = 1.0 \times 10^{13}$) corresponds to the direction of the densest system of closed fractures (running N-S). The values of the fracture stiffness differ practically by an order of magnitude.

In view of the spacing of the fractures observed under field conditions, one may assume that the laboratory spherical rock specimens (collected at the point of field measurement) do not contain these fractures. This is given by the condition of the solidness of the material used to produce the spherical specimen. The laboratory specimens thus do not represent the distribution and size of the fractures in the field, but the results of the measurements provide valuable information on wave propagation in a pseudo-continuous undisturbed rock medium. This affects the high values of the seismic-wave velocities observed during laboratory measurements.

If laboratory data are not available, the value of the maximum velocity, observed in the field, may be substituted into (11.5) for velocity V in the undisturbed medium. In this simplified case, it may be assumed that the anisotropy is caused only by the system of fractures running E-W, and that the other two fracture systems are too stiff to affect the wave propagation. The minimum observed velocity, corresponding to the observed anisotropy of 25% is substituted for group velocity U. In this case, for the E-W fracture stiffness, we obtain the value $\kappa = 1.55 \times 10^{12}$ Pa/m (Fig. 11.2, Curve B) which agrees with the results reported in Worthington's paper (2008).

The difference between the fracture stiffness values, determined with and without including the result of the laboratory velocity measurements for the system of open E-W fractures, is as much as 20%. In view of the scatter of the fracture stiffness estimates (Worthington, 2008), however, this difference should not be considered significant.

The fabric of the undisturbed rock in the outcrop plane is omni-directional, free of micro disturbances, which perhaps means that no directions of preferred deformation, observed in other cases, exist in it (Přikryl et al., 2004). In such a case, the directions of macroscopically observable systems of cracks could correspond fully to the orientation of the stress field, which was responsible for its generation. This would then provide considerable support to the studies of the stress field, which took part in the dislodgement of the whole geological body, or which acted on it in the geological past. This hypothesis can also be verified by laboratory loading tests of oriented rock specimens.

11.5   Conclusions

Field seismic measurements were carried out in the rock massif disturbed by three systems of fractures. Found out anisotropy of seismic velocity was analysed applying displacement discontinuity method. For calculation of fracture stiffness, laboratory measurements were employed. It was considered that the laboratory rock sample represents undisturbed homogenous material. Laboratory velocity anisotropy was practically negligible, while field velocity anisotropy (measured in the same rocks) was high. The combination of the field and laboratory velocity data was successfully used for the fracture stiffness evaluation. Obtained results are real and they agree with stiffness values reported in previous papers by other authors.

It was shown that the laboratory measured velocity values are higher than the field ones. The obtained results indicate the difficulty of direct transferring of the laboratory velocity values to the field conditions.

The described measurement methods (combination of laboratory and field) and data processing can provide considerable support to the studies of the stress field which took part in the dislodgement of the whole geological body, or which acted on it in the geological past.

Acknowledgments This work was partially supported by the Grant Agency of the Academy of Sciences of the Czech Republic, project No. IAA300130906, Research Project of Ministry of Education No. MSM 0021620855 and by Institutional Research Plan No. AVOZ 30130516 of Institute of Geology AS CR v.v.i.

References

Musgrave M.J.P. (1970) *Crystal acoustic*, San Francisco, Holden Day.
Pros Z., and Podroušková Z. (1974) *Apparatus for Investigating the Elastic Anisotropy on Spherical Samples at High Pressure*, Veröffentlichungen des Zentralinstituts für Physik der Erde **22**, 42–47.

Pros Z. (1977) *Investigation of anisotropy of elastic properties of rocks on spherical samples at high hydrostatic pressure*, In: High pressure and temperature studies of physical properties of rocks and minerals, Naukova Dumka, Kyjev, 56–57 (in Russian).

Přikryl R., Vilhelm J., Lokajíček T., Pros Z. and Klíma K. (2004) *Correlation of field seismic refraction data with 3-D laboratory ultrasonic sounding data during exploration of a dimension stone deposit*, Journal of Applied Geophysics **56**, 59–72.

Pyrak-Nolte L.J., Myer L.R. and Cook N.G.W. (1990) *Anisotropy in Seismic Velocities and Amplitudes From Multiple Parallel Fractures*, Journal of Geophysical Research **95**, B7, 11345–11358.

Worthington M.H. (2008) *Interpreting seismic anisotropy in fractured reservoirs*, First Break **26**, 57–63.

Chapter 12
Electrical Resistivity Imaging Method in Investigating Post-Glacial Geomorphological Forms

Anna Wysowska and Jolanta Pierwoła

Abstract The survey was carried out in the area of the Szczecinek Lakeland, which is part of the Southern Pomeranian Lake District mesoregion. The research area was situated about 1 km north of the village of Kądzielnia. The aim of the study was to investigate the geological structure of post-glacial sediments using electrical resisitivity imaging. The measurements were taken along three profiles, two of them with the Wenner array and one with the Schlumberger array. For each geoelectrical cross-section, four layers were distinguished. The average value of electrical resistivity for the first layer was about 2,000 Ωm, for the second layer about 800 Ωm, for the third layer about 200 Ωm and for the forth layer it was about 40 Ωm. The highest-resistivity fifth layer (of about 3,000 Ωm) was measured on profile 2. Determining the thickness of each layer was difficult because of their complicated geological structure, which was a result of fast and intense glacial movements. It was possible to distinguish parts belonging to the moraine or valley.

12.1 Introduction

The aim of study was to investigate the geological structure of post-glacial sediments using the geoelectrical method. The measurements were taken in the area of the Szczecinek Lakeland, which is a part of the Southern Pomeranian Lake District (Fig. 12.1; Kondracki, 2001; Karczewski, 1997). Characteristic features of the geomorphological structures which were formed during the last glaciation are high amplitudes and considerable thicknesses of sediments, ranging from 80 to 150 m (Dobracki and Lewandowski, 2002).

In the past, the geomorphology of the above-mentioned area was examined and described in detail. In the literature it is possible to find some works based on borehole analyses, which were carried out in this area. Classical geological and

A. Wysowska • J. Pierwoła (✉)
Department of Applied Geology, University of Silesia, Będzińska 60, 41 200 Sosnowiec, Poland
e-mail: jolanta.pierwola@us.edu.pl

A.F. Idziak and R. Dubiel (eds.), *Geophysics in Mining and Environmental Protection*,   117
Geoplanet: Earth and Planetary Sciences 2, DOI 10.1007/978-3-642-19097-1_12,

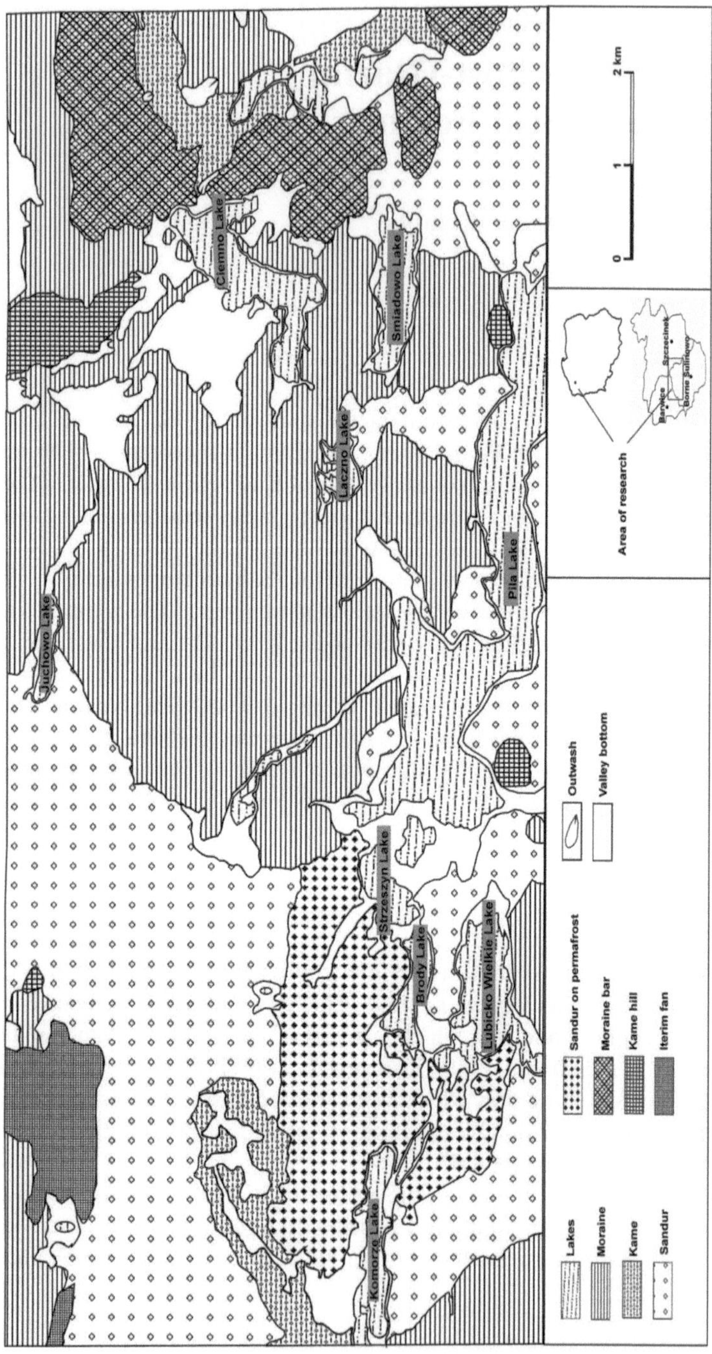

Fig. 12.1 Studied area (according to Karczewski, 1997)

geomorphological methods have primarily been applied to identify post-glacial structures (Klimek and Lewandowski, 2002).

12.2 Methodology of the Geophysical Investigation

The geophysical methods have recently become a common and useful tool in geomorphological studies. They include vertical electrical sounding and resistivity imaging (Etzelmüller et al., 2003; Schmitt et al., 2004), ground penetrating radar (Berthling et al., 2000) and seismic methods (King and Garg, 1980). The electrical properties of sediments provide information about the composition and shape of the subsurface layers in a non-destructive way. Beginning in the 1960s, the vertical resistivity soundings were applied to get information about glacial sediments in the Arctic (Wagner, 1996), Antarctica (López-Martinez and Serrano-Cañadas, 2005) and in mountain areas (Kneisel et al., 2000; Etzelmüller et al., 2003). It was possible to determine the structure and thickness of sediments or the permafrost frontier (Ishikawa and Hirakawa, 2000; Etzelmüller et al., 2003; Hauck and Vonder Mühll, 2003; Kneisel and Hauck, 2003). Resistivity imaging combines features of resistivity sounding and profiling methods. This method allows ground structure disorders to be investigated to determine subsurface transport properties and monitor resistivity over a long-term (Olsen et al., 1999).

The inversion process is essential for this method (Loke, 2004; Rudzki, 2002). The apparent resistivity can be described using the mathematical function (Samouëlian et al., 2005) associated with the properties of the sediments:

$$d_i = F_i(m_1, m_2, ..., m_m) + e_i \qquad (12.1)$$

where d_i is the observed data ($i = 1, ..., n$), F_i is the forward mapping, m_j are the model parameters, and e_i is the RMS error.

A comparison of the theoretical model with field measurements relies on calculating the electrical resistivity at individual measuring points (Sharma, 1997). The start model is related to presumable values of the resistivity m_j through theoretical c_i values:

$$c_i = F_i(m_1, m_2, ... m_m) \qquad (12.2)$$

The RMS error is estimated on the basis of the difference between the observed and theoretical values:

$$e_i = d_i - c_i \qquad (12.3)$$

Finding a set of data with minimum differences between the observed and theoretical data is the aim of the inversion process (Samouëlian et al., 2005). Data is fitted using the least squares method (de Groot-Hedin and Constable,

1990). The adjustment is modified in the following iterations according to the equation:

$$(JJ' + uQ)m = J'd \qquad (12.4)$$

where: $Q = q_x q_{x'} + q_z q_{z'}$, d is the discrepancy vector, m is the model perturbation vector, u is the damping factor, J is the matrix of partial derivatives, q_x is the horizontal flatness filter, and q_z is the vertical flatness filter.

The damping factor and flatness filters, which can be adapted to different data types, are advantages of this method (Sjödahl, 2006).

12.3 The Area of Investigations

The total thickness of the Pliocene sediments in the studied area is about 175 m (Bukowska-Jania and Pulina, 1997; Geological Map of Poland 1:50,000). The border between the Pliocene and Pleistocene is situated at a depth of 100 m on average (Karczewski, 1991, 1997; Dobracki and Lewandowski, 2002; Klimek and Lewandowski, 2002).

Boulder clays, sands, water-glacial gravels and silts, as well as lake-sands and peats originating from the interglacial periods are typical Pleistocene sediments of the studied area (Bukowska-Jania and Pulina, 1997). The geological and hydrogeological structure was measured using a dozen or so boreholes. A shallow subsurface is built of sands and gravels of different granulations. The groundwater level is situated at depths of 8–14 m. A boulder clays level is situated beneath the above-mentioned sediments (Sołtysik and Rybka, 1993).

Two moraine-bands related to the Krajno glaciation phase and the occurrence of numerous small post-glacial lakes are characteristic for the studied lakeland. Close to 7% of the region is covered by surface water. The large retention capacity is caused by the number and volume of water reservoirs, peat-bogs and the existence of large areas with sandy and gravel sediments (Dynowska, 1971). The chemical composition of the water is characterised by average mineralization (260–300 mg/l) and the presence of the calcium carbonate ($CaCO_3$), which leaches from the glacial and water-glacial sediments.

The research area is located 1 km north of the village of Kądzielnia. Three parallel measurement profiles were traced on the northern hillside. The orientation of the survey lines was NNW-SSE and changes in the level did not exceed 8 m. They crossed two morphological structures: a moraine and a glacial valley. A moraine is a hill or a band of hills built of material which melted from the glacier or was displaced under its pressure. It is composed of unconsolidated sediments with different granulations up to large boulders. Glacial valleys were formed by ice bodies or water which flowed from the melting glacier. The so-called boulder clays were deposited on the bottom of the valleys.

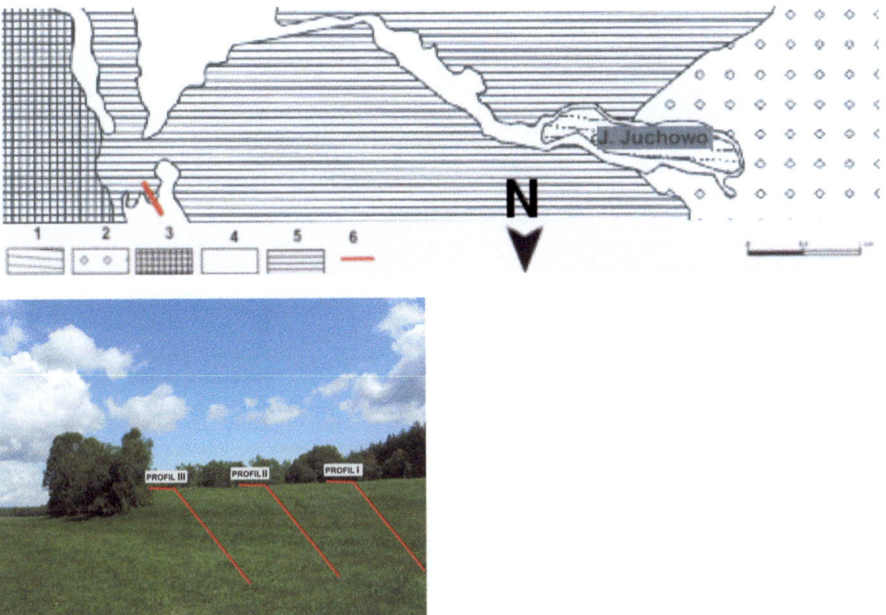

Fig. 12.2 Location of the research lines on the basis of a geomorphological map. *1* – lake, *2* – sandur, *3* – kame, *4* – river valley, *5* – moraine, *6* – resistivity imaging profiles (according to Karczewski, 1997)

The Wenner (profiles 1 and 3) and the Schlumberger (profile 2) arrays were used to take measurements. The distance between the profiles was 10 m. In Fig. 12.2 the exact location of the research traverses is presented. The length of profiles 1 and 3 was 100 m and the distance between electrodes was 2.5 m. Profile 2 was of 120 m in length and the electrode spacing was 3.5 m. The average depth of geoelectrical prospecting was 20 m. Measurements were performed using an ABEM LUND Imaging System. The applied current intensity varied in the range 20–200 mA. The obtained resistivity data were of satisfactory quality. The standard deviation for a particular data point did not exceed several percent. To interpret the measurement results RES2DINV software was used.

12.4 Results

The obtained resistivity cross-sections are presented in Figs. 12.3–12.5. Determining the thickness of a particular layer was difficult because of the disturbed geological structure of the sediments, which was a result of rapid and intense glacial movements. Average resistivity values for the investigated layers are presented in Table 12.1.

Because the survey lines were situated close to each other, within two neighboring structures, they are discussed together.

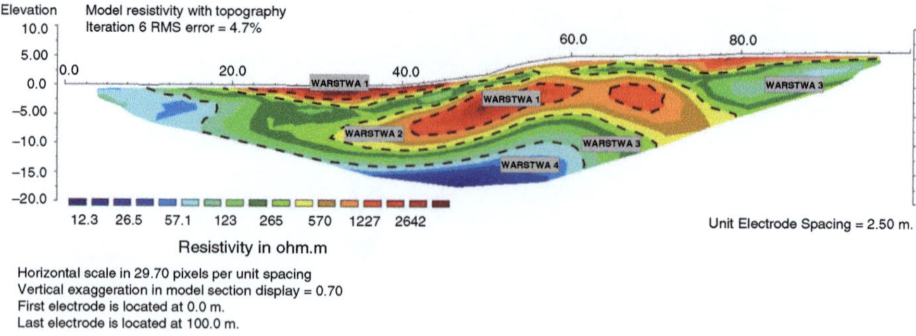

Fig. 12.3 Resistivity inversion model, profile 1

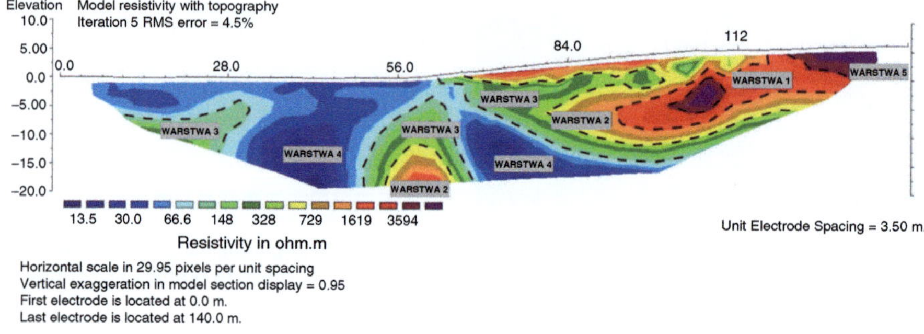

Fig. 12.4 Resistivity inversion model, profile 2

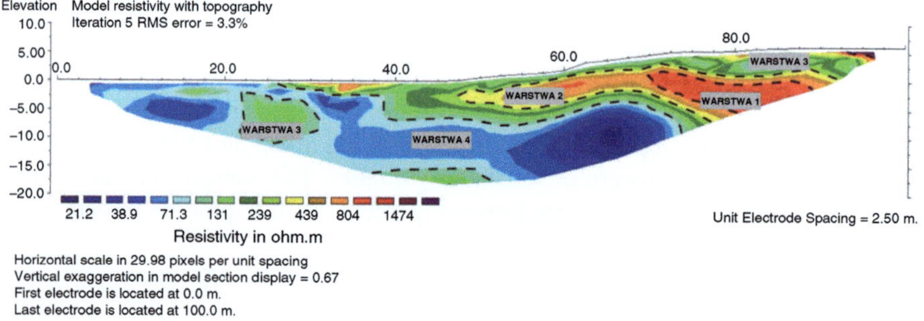

Fig. 12.5 Resistivity inversion model, profile 3

It is possible to identify layers 1–4 in all sections. The first layer consists of sediments with a resistivity value of about 1,500–2,000 Ωm. This can be interpreted as coarse-grained gravels and sands. The second layer is characterised by electrical resistivity ranging from 500 to 1,000 Ωm. These are probably sands and gravels with a higher content of the fine fraction. According to the position and the resistivity values it is possible to conclude that two top layers are dry. The above-mentioned sediments form thin isolated planes close to the surface and convexed

Table 12.1 Average values of electrical resistivity obtained for interpreted layers

Layer	Profile	Electrical resistivity [Ωm]
1	1	2,000
	2	1,700
	3	2,000
2	1	750
	2	750
	3	800
3	1	200
	2	200
	3	200
4	1	30
	2	45
	3	30
5	2	3,000

upward lenses with a thickness of several meters. Such forms are characteristic morainal structures and they are the result of intensive glacitectonical processes (overthrusts which are the result of squeeze) (Embleton and Thornes, 1985). It is possible for the upper part of these structures to be eroded.

Sediments with a similar resistivity value can be observed in the lower part of section 2 at the 60th m. It can be surmised that this is a large boulder which was deposited among the fine-grained sediments.

The observed resistivity values for the third layer vary between 100 and 300 Ωm. This layer is composed of fine-grained sands which are water-bearing at the bottom. These are valley sediments which are covered by moraine sediments. Their complicated structure is the result of intensive glaciotectonic processes.

The fourth layer is composed of sediments with resistivity from 10 to 50 Ωm. This layer consists of boulder clay forming the bottom of the valley. Clays are interbedded with fine-grained sands which are typical fluvioglacial deposits. The top of this layer is situated directly below the surface in the southern part of the sections; in the northern part it is situated considerably deeper and reaches a depth of 20 m. It is possible that clayey sediments belong to different glacial phases (Żogała et al., 2008).

Layer 5 was only identified in profile 2. This is a high-resistivity layer with a resistivity of above 3,000 Ωm. The presence of blocks of rock inside unconsolidated sediments can occur.

12.5   Conclusion

The electrical imaging method is a valuable tool for investigating shallow, postglacial sediments. This method enables the identification of the borders between different geomorphological forms. It was possible to distinguish parts which belong to the moraine and glacial valley for the investigated profiles.

Electrical imaging makes it possible to investigate structures with high accuracy. It can be useful for tracing the exact course of post-glacial structures in the areas between bore-holes and it can be a complement to logging information.

References

Berthling I., Etzelmüller B., Isaksen K., and Sollid J. L. (2000) The rock glaciers on Prins Karls Forland (II): GPR Soundings and the Development of Internal Structures, Permafrost and Periglacial Processes, 11, 357–369.

Bukowska-Jania E., and Pulina M. (1997) Problemy hydrochemiczne i kriochemiczne w rejonie Bornego Sulinowa (Pomorze Zachodnie) [Hydrochemical and kriochemical problems in the Borne Sulinowo region, West Pomerania], In: Kostrzewski A. (red.), Rzeźba i osady czwartorzędowe obszarów współczesnego i plejstoceńskiego zlodowacenia półkuli północnej, Wyd. UAM, Ser. Geografia 58, 29–49 (in Polish).

De Groot-Hedin C., Constable S. (1990) Occam's inversion to generate smooth, two dimentional models from magnetotelluric data, Geophysics 4, 113–123.

Dobracki R., and Lewandowski J. (2002) Plejstocen Pojezierza Drawskiego i Szczecineckiego [Pleistocene of Drawsko and Szczecin Lake Districts], In: Plejstocen Pomorza Środkowego i strefa marginalna lobu Parsęty – IX Konferencja: Stratygrafia Plejstocenu Polski. Państwowy Instytut Geologiczny, Oddział Pomorski, Szczecin i Wydział Nauk o Ziemi, Uniwersytet Śląski, Katowice (in Polish).

Dynowska I. (1971) Typy reżimów rzecznych w Polsce [Water regime types in Poland], Zeszyt Naukowy Uniwersytetu Jagielońskiego, Prace Geograficzne 28: 151 (in Polish)

Embleton C., and Thornes J. (1985) Geomorfologia dynamiczna [Dynamical Geomorphology], Państwowe Wydawnictwo Naukowe, Warszawa (in Polish),

Etzelmüller B., Berthling I., and Ødegård R.S. (2003) One-dimensional DC-resistivity depth sounding as a tool in permafrost investigations in high mountain areas of Southern Norway, in: L. Schrott, A. Hordt and R. Dikau (eds) Geophysical Applications in Geomorphology, 19–36.

Hauck C., and Vonder Mühll D. (2003) Evaluation of geophysical techniques for application in mountain permafrost studies, in: L., Schrott, A., Hoerdt, R., Dikau, Geophysical Methods in Geomorphology, Zeitschrift für Geomorphologie, Suppl. 132, 161–190.

Ishikawa M., and Hirakawa K. (2000) Mountain permafrost distribution based on BTS measurements and DC resistivity soundings in the Daisetsu Mountains, Hokkaido, Japan, Permafrost and Periglacial Processes, 11, 109–123.

Karczewski A. (1991) Rozwój i zasięg fazy pomorskiej lobu Parsęty podczas zlodowacenia vistuliańskiego [Development and range of the Pomeranian phase of the Parseta lobe during Vistulian glaciation], In: Kostrzewski A., (ed.) Geneza, litologia i stratygrafia utworów czwartorzędowych, Geografia 50, Wydawnictwo Naukowe Uniwersytetu im. A. Mickiewicza, Poznań (in Polish).

Karczewski A. (1997) Geomorfologia i rozwój rzeźby przedpola fazy pomorskiej w północnym sąsiedztwie jeziora Pile [Geomorphology and development of the Pomeranian phase forefield in the northern vicinity of the Pile lake], In: Bukowska-Jania E. and Pulina M., (eds.), Studia nad środowiskiem geograficznym Bornego Sulinowa, PWN, Warszawa (in Polish).

King M.S., and Garg O.P. (1980) Interpretation of seismic and resistivity measurements in permafrost in Northern Quebec, Proceedings, 5th Symposium on Permafrost Geophysics, November 1978, National Research Council of Canada, Ottawa, Technical Memorandum, 128, 50–69.

Klimek K., and Lewandowski J. (2002) Stanowisko nr 5 – Liszkowo, In: Plejstocen Pomorza Środkowego i strefa marginalna lobu Parsęty – IX Konferencja: Stratygrafia Plejstocenu

Polski, Państwowy Instytut Geologiczny, Oddział Pomorski, Szczecin i Wydział Nauk o Ziemi, Uniwersytet Śląski, Katowice (in Polish).

Kneisel C., Haeberli W., and Baumhauer R. (2000) Comparison of spatial modelling and field evidence of glacier/permafrost relations in an alpine permafrost environment, Annals of Glaciology 31, 269–274.

Kneisel C., and Hauck C. (2003) Multi-method geophysical investigation of a sporadic permafrost occurrence, in: Geophysical Methods in Geomorphology, L. L., Schrott A., Hoerdt R., Dikau Zeitschrift für Geomorphologie, Suppl. 132: 145–159.

Kondracki J. (2001) Geografia Polski. Mezoregiony fizyczno geograficzne [Geography of Poland], Państwowe Wydawnictwo Naukowe, Warszawa (in Polish).

Loke M.H. (2004) Tutorial: 2-D and 3-D electrical imaging surveys in: www.geoelectrical.com.

López-Martinez J., and Serrano-Cañadas E. (2005) Mapping of permafrost evidence in the South Shetland Islands, Western Antarctica, in: Dr. James G. Bockheim International Workshop on Antarctic Permafrost and Soils, Office of Polar Programs, Antarctic Section, 69, University of Wisconsin, Madison.

Olsen P.A., Binley A., Henry-Poulter S., and Tych W. (1999) Characterizing solute transport in undisturbed soil cores using electrical and X-ray tomographic methods, Hydrol. Process. 13, 211–221.

Rudzki M. (2002) Zastosowanie metody tomografii elektrooporowej do wykrywania podziemnych obiektów antropogenicznych [The use of electrical resistivity method to detect subsurface antropogenic objects], Publ. Inst. Geophys. Pol. Acad. Sci., M-25, 313 (in Polish).

Samouëlian A., Cousin I., Tabbagh A., Bruand A., and Richard G. (2005) Electrical Resistivity survey in soil science: a review. In: Soil & Tillage Research 83, 173–193.

Schmitt G.A., Shindell D.T., Miller R.L., Mann M.E. and Rind D. (2004) General circulation modelling of Holocene climate variability, Quaternary Sci. Rev., 23, 2167–2181

Sharma P.V. (1997) Environmental and Engineering Geophysics, Cambridge University Press, Cambridge.

Sjödahl P. (2006) Resistivity investigation and monitoring for detection of internal erosion and anomalous seepage in embankment dams, Doctoral Thesis, Lund University, Lund.

Sołtysik J., and Rybka A. (1993) Raport o zanieczyszczeniach gruntu i wód podziemnych produktami ropopochodnymi i innymi substancjami chemicznymi na obiekcie opuszczonym przez wojska Federacji Rosyjskiej Borne Sulinowo [Report on soil and water contamination by chemical substances in Borne Sulinowo], Wojskowa Akademia Techniczna, Exbud – 22 Hydrogeotechnika, Warszawa, Kielce (in Polish).

Wagner S. (1996) DC resistivity and seismic refraction soundings on rock glacier permafrost in northwestern Svalbard, Norsk Geografisk Tidsskrift, 50, 25–36.

Żogała B., Dubiel R., Lewanowski J., Zuberek W., and Gąska G. (2008) Application of resistivity imaging method for investigation of geologic structure of Pleistocene sediments, Acta Geodyn. Geomater., V.5, No.2, 177–183.

Chapter 13
Frictional Metamorphism of Coal in the Upper Silesia Coal Basin, Southern Poland

Stanisław Roman Ćmiel and Maria Dziurowicz

Abstract Measurements of parameters of coal in the fault zones of the Upper Silesia Coal Basin (USCB) revealed that ca. 12% of the examined samples showed aggrading changes of coal quality; ca. 30% of them showed degrading ones and 58% did not show any of these changes. The aggrading changes were observed only in the immediate vicinity of the slickenside at a maximum distance of 0.2 m. The noticeable changes were connected with the majority of the parameters tested. The mean values of the coefficient of variation for all of the tested parameters ranged from 1 to 100%; the average value was 16.6%, and changeability compared to the unchanged zone varied from 0 to 51.5%. These changes are connected with flexible faults whose run is close to N-S and W-E directions.

13.1 Introduction

In many parts of the Upper Silesia Coal Basin (USCB), the secondarily altered coal characterized by increased or reduced quality of coal parameters has been reported. Typical examples of degrading changes in coal formations are the so-called red beds, which occur in outcrops of coal seams. Similar, but not as intensive changes can be observed in fault zones. Qualitative testing conducted by the authors in the fault zones of the USCB revealed that ca. 12% of the samples examined showed aggrading changes of coal quality, ca. 30% of them showed degrading ones, and 58% did not show any of these changes. Measurements were carried out at an interval of 1–10 m from the fault plane; however, the aggrading changes were observed only in the immediate vicinity of the slickenside, at a maximum distance of 0.2 m. Noticeable changes were connected with the majority of the parameters tested (Ćmiel, 2009).

S.R. Ćmiel (✉) • M. Dziurowicz
University of Silesia, Będzińska 60, 41–200 Sosnowiec, Poland
e-mail: stanislaw.cmiel@us.edu.pl

A.F. Idziak and R. Dubiel (eds.), *Geophysics in Mining and Environmental Protection*,
Geoplanet: Earth and Planetary Sciences 2, DOI 10.1007/978-3-642-19097-1_13,
© Springer-Verlag Berlin Heidelberg 2011

The mean values of the coefficient of variation for all of the parameters tested of this type ranged from 1 to 100%; the average value was 16.6% and changeability compared to the unchanged zone varied from 0 to 51.5%. In the examined faults, an analysis of the direction and changeability of coal quality in comparison to the geometric parameters, such as a dip of the azimuth of the fault plane, a throw of the fault and its depths, showed a noticeable regularity.

13.2   Test Results

Table 13.1 shows an example of changeability for individual parameters of coal in the case of aggrading changes based on the sample which was taken from the P-3 cross-cut belonging to the 405 coal seam of the Wujek Coal Mine. The specimen for analysis was collected at a depth of 613 m in the vicinity of the Kłodnica fault at an interval of 1–10 m. In this area, the throw of the fault is ca. 160 m, its dip azimuth is 180° and the dip of the fault plane is 80°.

Among the technical parameters analyzed, only the calorific value increases near the fault plane, whereas the other parameters decrease. The average moisture in the fault zone is 2.90% and on the surface of the fault the minimum value is 2.05%. In comparison with the mean value for the unchanged zone, the moisture falls by 31.9% with an average coefficient of variation of 22.4%. On the surface of the fault, the mean value of the ash content is 10.0% and in comparison with the average value for the unchanged zone the increase of this parameter reaches 15.8% for a low coefficient of variation of 15%. The mean content of the volatile matter near the fault zone is 28.2% and on the surface of the fault it decreases up to 27.5%; in comparison with the mean value for the unchanged zone, the content of this parameter falls on average by 19.4% at the low value of the coefficient of variation 8.7%. The highest calorific value can be found on the surface of the fault, at an average of 34.4 MJ/kg for this zone. In comparison with the average value for the unchanged zone, this parameter increases on average by 6.2% at the low value of the coefficient of variation (2.9%).

Parameters of the elementary analysis vary; the contents of some elements such as carbon (C) and sulphur (S) increase, whereas the oxygen content decreases, along with hydrogen (H) and nitrogen (N) content, which also decrease but to a lesser degree. The content of nitrogen found near the fault zone, which is 1.42% on average, falls slightly, by 3.3%, in comparison with the unchanged zone at the low value of the coefficient of variation (2.8%). Similarly, in the immediate proximity of the fault, the oxygen content is 7.5% on average, i.e., it falls by 19.0% in comparison with the mean value for the unchanged zone at the low value of the coefficient of variation (10.4%).

The carbon content (C) found on the surface of the slickenside is 86.35% on average, i.e., it increases slightly, by 3.1%, in comparison with the mean value for the unchanged zone at the low value of the coefficient of variation (1.5%).

Table 13.1 The increase in the rank of coal on the surface of the fault

Distance [m] parameters	0	0.2	0.4	0.6	0.8	1	2	5	8	10	Average 0–10 m	Average 0–0.2 m	Average 0.4–10 m	Standard deviation	Coefficient of variation	Variation [%]
M^a [%]	2.05	2.16	2.21	2.36	2.92	3.38	3.12	3.54	3.62	3.65	2.90	2.11	3.10	0.65	22.4	−31.9
A^a [%]	10.6	9.4	8.9	9.1	9.2	9.5	7.6	8.8	8.2	7.7	8.91	10.00	8.63	1.34	15.0	15.8
V^{daf} [%]	27.5	28.9	33.8	34.5	34.7	35.2	35.8	35.6	34.9	35.4	33.63	28.20	34.99	2.94	8.7	−19.4
CV^{daf} [MJ/ kg]	34.5	34.3	32.5	32.7	32.9	31.6	32.4	32.2	32.8	31.9	32.78	34.40	32.38	0.94	2.9	6.2
N^{daf} [%]	1.41	1.42	1.45	1.51	1.46	1.45	1.50	1.42	1.49	1.43	1.45	1.42	1.46	0.04	2.8	−3.3
O^{daf} [%]	7.3	7.7	8.7	8.3	8.8	9.3	10.1	9.7	9.9	9.3	8.91	7.50	9.26	0.93	10.4	−19.0
C^{daf} [%]	86.5	86.2	84.7	84.2	83.9	83.8	82.7	82.9	84.2	83.4	84.25	86.35	83.73	1.26	1.5	3.1
H^{daf} [%]	4.80	4.60	4.60	4.90	5.10	5.00	4.74	4.81	4.78	4.86	4.82	4.70	4.85	0.16	3.3	−3.1
S_t^{a} [%]	0.87	0.89	0.84	0.91	0.95	0.73	0.97	0.87	0.81	0.89	0.87	0.88	0.87	0.07	8.0	1.0
S_p^{a} [%]	0.67	0.65	0.65	0.65	0.71	0.58	0.63	0.69	0.57	0.62	0.64	0.66	0.64	0.04	6.2	3.5
S_{so4}^{a} [%]	0.01	0.01	0.02	0.01	0.01	0.03	0.01	0.00	0.01	0.00	0.01	0.01	0.01	0.01	90.9	−11.1
RI	27	22	19	15	20	16	18	15	17	19	18.80	24.50	17.38	3.65	19.4	41.0
a [%]	28.3	23.7	21.5	17.4	22.1	12.8	16.2	16.6	20.4	13.0	19.20	26.00	17.50	4.91	25.6	48.6
R_m^{a} [%]	0.85	0.81	0.77	0.75	0.78	0.75	0.73	0.77	0.75	0.78	0.77	0.83	0.76	0.03	3.9	9.2
d_r [Mg/m^3]	1.49	1.47	1.42	1.39	1.41	1.36	1.42	1.38	1.41	1.39	1.41	1.48	1.40	0.04	2.8	5.9

The hydrogen content (H) in the fault zone is 4.7% on average and falls by 3.1% in comparison with the mean value for the unchanged zone. There is little variability of the hydrogen content, which is reflected in the low value of the coefficient of variation (3.3%).

The mean value of the sulphur content changes slightly; the contents of total sulphur and pyrite sulphur tend to increase, whereas the relative contribution of sulphate sulphur does not show any distinct changes. In the fault zone the mean contribution of total sulphur is 0.88% and increases by 1.0%, and pyrite sulphur is 0.66% and falls slightly, by 3.5%, at the low values of the coefficient of variation, of 8.0 and 6.2%, respectively. Sulphate sulphur occurs in minute quantities.

Coking properties measured in the fault zone show considerable improvement; however, the coal tested generally shows poor coking ability. The mean values of Roga's Index and contraction in the area of the slickenside are 24.5 and 26%, respectively. The values of these parameters increase significantly in comparison with the averages for the unchanged zone, by 41 and 48.6%, respectively, reaching the mean values of the coefficient of variation of 19.4 and 25.6%, respectively.

The values of physical factors tested in the area of the fault plane show moderate variability. The mean value of vitrinite reflectance reaches 0.83% and in comparison to the average for the unchanged zone it increases by 9.2% in the area of the slickenside at the low value of the coefficient of variation (3.9%). The reflectograms, which are used to characterize the rank of coal, in the slickenside zone are of flattened and asymmetric types, and generally they are extended in the direction of diminishing values (minus on the skew) in comparison with the reflectograms for the unchanged zone (Fig. 13.1). The mean value of coal density amounts to 1.48 Mg/m^3, and on the surface of the fault it increases insignificantly,

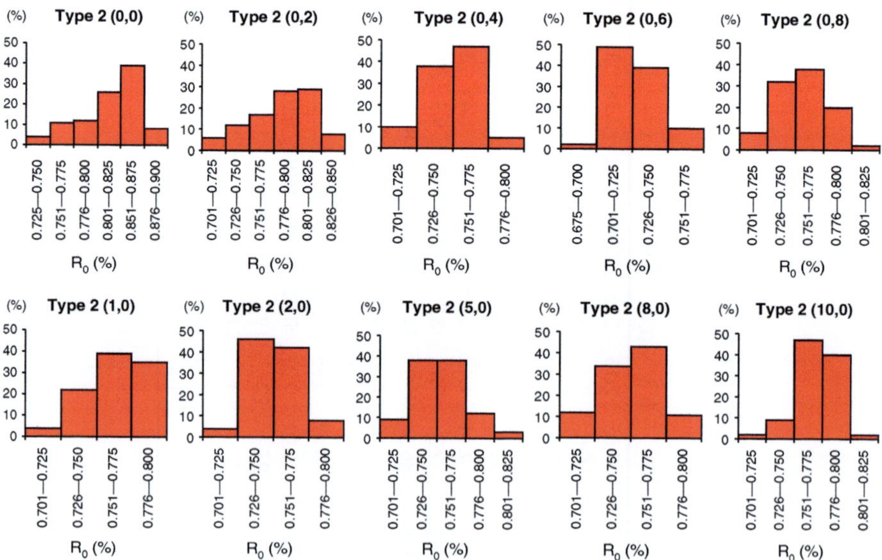

Fig. 13.1 Reflectograms of aggrading changes of coal quality in the fault zones

by 5.9%, in comparison with the average for the remaining unaltered zone at the low value of the coefficient of variation (2.8%).

13.3 Discussion

The distance and scale of aggrading changes in coal parameter values found in fault zones are definitely lower in comparison with hypergenic changes, whereas their opposite direction of change results in an increase in the rank of coal. The changes in the values of coal parameters are observed directly only on the surface of the fault.

The faults with the aggrading coal change of in fault zones run in a different direction compared to the others, by running close to the N-S and E-W directions (Fig. 13.2a–c). According to Herbich (1981), there are approximately 80% of the

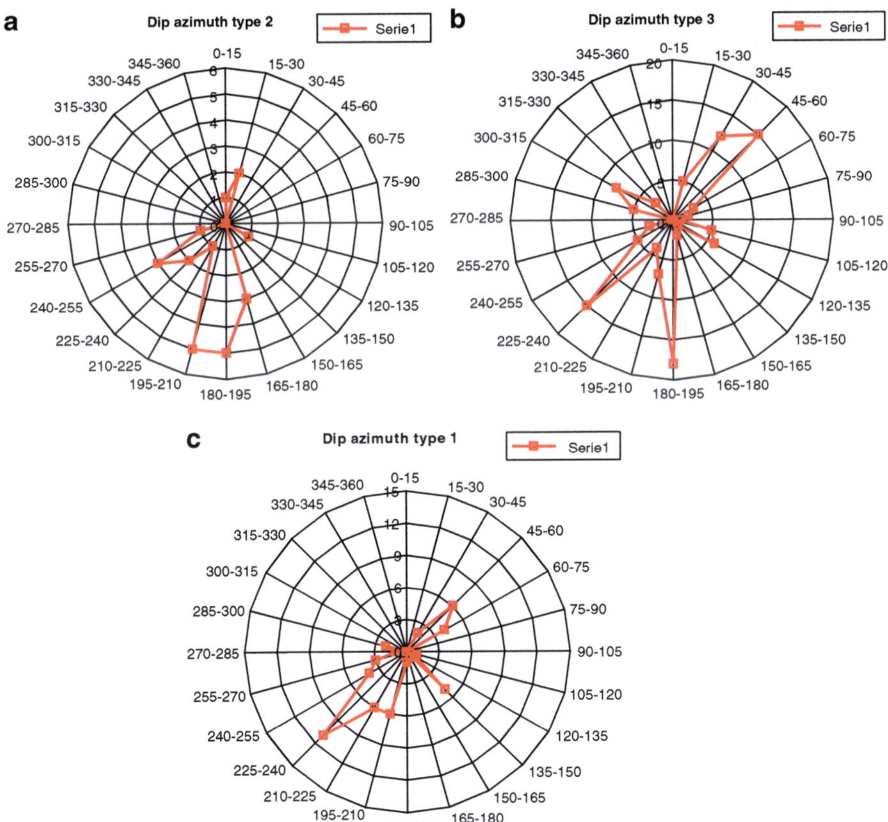

Fig. 13.2 (**a**) Dip azimuth chart for aggrading of coal quality in the fault zones. (**b**) Dip azimuth chart for coal quality in the unchanging fault zones. (**c**) Dip azimuth chart for weathering changes in the fault zones

undulating structure faults showing such directions in the Upper Silesia Coal Basin. They are formed as a result of the slow and gradual propagation of shear movement. That is why they are so flexible. Such faults form in a compression regime in the final phase of elastic strain when the value of this strain exceeds the flexibility of the rock in particular conditions. In the vicinity of faults running N-S and E-W one can also observe an increase of gas pressure, which proves their closed nature. Destruction of the rock occurs in an evolutionary way through the gradual concentration of originally dispersed slips in zones which were subject to deformation (Jaroszewski, 1994).

Shearing stress increased gradually and this process simultaneously generated the heat of friction, whereas the concentration of slips as well as the closing of a fault fissure created favourable conditions for the accumulation of heat which finally increased the rank of coal in seams. That was why the migration of gas as well as mineralized solutions was limited. Taking this into consideration, the process can be defined as the friction metamorphism of coal (Scholz et al., 1979). Larger fragments and breccia of surrounding rocks as well as sulphide and carbonate association have not been found in the fissures formed in these faults, which is confirmed by scientific research carried out on the formation of the Kłodnica fault (Nizicki et al., 2004). However, the distinction of brittle-type and flexible faults is hindered because mixed types of faults mostly occur. It can be assumed that the changes which resulted in increasing degree of coalification took place in the fault zones with a predominant deformation component of a flexible type.

13.4 Summary

The analysis carried out has shown that aggrading changes of coal quality occur only in the immediate vicinity of the fault plane. These changes are connected with flexible faults whose run is close to NE and WE directions. The origin of observed changes is related to the frictional heat generated on the fault plane causing frictional metamorphism of coal. The process of closing the fault fissure favored the accumulation of heat causing the reduction of permeability to gases and mineralized solutions simultaneously.

Acknowledgements The study was financially supported by Commissioned Research Project KBN PB-W-04-068-00-09 from the Polish funds for science.

References

Ćmiel S.R. (2009) Charakterystyka epigenetycznych zmian węgla w pokładach w strefach uskokowych GZW [Characterization of epigenetic changes of coal fault zones of the Upper Silesia Coal Basin]. Prac. Nauk. UŚ. nr. 2682, 109 pp (in Polish).

Herbich E. (1981): Analiza tektoniczna sieci uskokowej GZW [Tectonic analysis of fault network of the Upper Silesia Coal Basin]. Rocz. Pol. Tow. Geol. 51, 3–4, 383–434 (in Polish).

Jaroszewski W. (1994) Uskoki i zjawiska pokrewne [Faults and related phenomena], in: Dadlez R., Jaroszewski, Tektonika. PWN. Warszawa, 88–162 (in Polish).

Nizicki R., Paszek D., and Borówka B. (2004) Przebieg i wykształcenie strefy uskokowej uskoku kłodnickiego w obszarze górniczym KWK Makoszowy [The Kłodnica fault zone in the mining region of Makoszowy Colliery], In: Mat. XXVII Symp. Geol. Form. Węglonośnych Polski. AGH, Kraków, 121–126 (in Polish).

Scholz C.H., Beavan J., and Hanks T.C. (1979) Frictional metamorphism, argon depletion, and tectonic stress on the Alpine fault, New Zealand. Journal of Geophysical Research, Volume 84, Issue B12, 6770–6782.

Index

A.F. Idziak and R. Dubiel (eds.), *Geophysics in Mining and Environmental Protection,*   135
Geoplanet: Earth and Planetary Sciences 2, DOI 10.1007/978-3-642-19097-1,
© Springer-Verlag Berlin Heidelberg 2011